# GETTY ON GETTY

Surely every man walketh in a vain shew:
surely they are disquieted in vain: he heapeth
up riches, and knoweth not who shall gather them.
                              Psalms XXXIX, 6

# GETTY ON GETTY

## A MAN IN A BILLION

Somerset de Chair

CASSELL

Cassell Publishers Ltd
Artillery House, Artillery Row, London, SW1P 1RT

Copyright © Somerset de Chair 1989

All rights reserved. No part of this publication may be reproduced or transmitted in any form or by any means, electronic or mechanical, including photocopying, recording or any information storage or retrieval system, without prior permission in writing from the publishers.

First published 1989

Distributed in the United States by
Sterling Publishing Co. Inc.
387 Park Avenue South, New York, NY 10016–8810

Distributed in Australia by
Capricorn Link (Australia) Pty Ltd
PO Box 665, Lane Grove, NSW 2066

**British Library Cataloguing in Publication Data**
De Chair, Somerset, *1911*–
    Getty on Getty: a man in a billion.
    1. United States. Petroleum industries. Getty, J.
    Paul (Jean Paul), 1892–1976
    I. Title
    338.7′6223382′0924

ISBN 0–304–31807–8

Phototypeset by Input Typesetting Ltd, London
Printed in Great Britain by
Biddles Ltd, Guildford and King's Lynn

# Contents

| | |
|---|---|
| Introduction | 1 |
| 1 Kidnappings, Hearst and Houses | 4 |
| 2 Businesses and Businessmen | 17 |
| 3 Art and the Malibu Museum | 37 |
| 4 The Early Days in Oil | 56 |
| 5 Romance and Marriage | 65 |
| 6 Famous Friends | 77 |
| 7 Early Travels | 87 |
| 8 Getty on Getty | 97 |
| 9 War | 109 |
| 10 Insights | 114 |
| 11 Personal Papers | 122 |
|     Getty's Unpublished Diaries | 122 |
|     Letters Getty Treasured | 138 |
| 12 Recollections of a Luncheon | 142 |
| 13 Conclusion | 148 |
| Appendix I | 161 |
| Appendix II | 163 |
| Index | 165 |

*For Juliet*

# INTRODUCTION

This is not a conventional biography of J. Paul Getty, oil magnate and philanthropist; it is based on conversations between two old friends, with part of it devoted to highlights of Paul Getty's comments on certain subjects, and some of his personal papers. However, it is a biography in the sense that it brings out Paul's personality and interests vividly. As Lansing Hays, his attorney, has said: 'It is the first book about Paul I have read in which he appeared as a fine man, and the first in which I truly recognized Getty, because you let him talk!' And it was approved by Getty himself and his New York attorney, as stipulated in his letter of 4 October 1973, authorizing the book.

The conversations took place between Paul and myself from October 1973 to April 1974. They were taped in the drawing-room at Sutton Place, Paul's headquarters near Guildford in Surrey, England, with his complete knowledge and approval. Indeed, as he stated: 'After the extraordinary business of the faked biography of Howard Hughes by Clifford Irving, I feel it desirable to state unequivocally that I have read and approved every word of the present book. I am quite prepared to be judged by it.' A complete transcript of the tapes themselves has been donated to the Columbia University Oral History Collection in New York City.

Paul, eighty-one at that time, always dressed soberly for our talks, rather in the London fashions than in country gentleman tweeds. His face was long and thin, with a fine nose and a high forehead, framed with wisps of brown hair. The penetrating blue eyes were the focal point of this famous face. His voice was measured, precise, and his words came slowly, due to recent illness. His mind was as hard and sharp and brilliant as a diamond, but the setting had worn a little loose over the years. He was the richest single individual to amass a fortune by his own efforts, and it was from Sutton Place, built by a friend of Henry VIII in 1521, that he controlled his companies.

The environment itself was comfortable. The drawing-room had been a kitchen in Tudor times and featured a large fireplace, kept lit to ward off the winter chills. The panelling, once stained in dark Victorian, was stripped back to its original light Honduras mahogany. Seventeenth-century Dutch paintings hung on the walls; the furniture was upholstered in pale lime green.

The problem in approaching the life of Paul Getty is that the whole is much greater than the parts. Nothing much seems to have happened in his life except success. He did not lead, by some standards, a very adventurous life, apart from his business activities. No; on the whole, he led a quiet life. What was astonishing about Paul Getty was his tremendous knowledge of art and works of art, allied to a business brain capable of making the millions necessary to acquire some of the finest objects in the world. It may well be that he was the most successful operator in the business arena since the capitalist system developed in the nineteenth century. What is interesting about him is not only how he made his money, but what he did with it; what sort of man he became as a result of earning it. Getty was as straight as a die, and there are businessmen all over the world who would tell you that they accepted his word on the telephone as a bond to honour any contract he agreed to make.

He possessed an almost old-world standard of courtesy. Paul invariably showed me to the front door at Sutton and unbolted it himself rather than leave it to a butler. He was becomingly modest; although he enjoyed publicity, he did not seek it. He was a quiet and unassuming man who nevertheless enjoyed the company of significant people. His Methodist upbringing had left him with a strong puritanical streak.

This book represents the only time, and certainly the last time, Paul talked about his life for posterity in taped form. My own impression of Getty was that he remained a sphinxlike and somewhat inscrutable character; but I shall leave it to the readers to judge for themselves.

I should like to acknowledge the courtesy and help I received from Getty's principal aides in the preparation of this book; notably his accountant, Norris Bramlett, his attorney, the late Lansing Hays, his principal private secretary, Barbara Wallace, and his secretary Carol Tier who arranged the meetings. Also Gael Eltringham of Connecticut for her professional help in condensing the original typescript of over 470 pages to its present length, and isolating the various subjects I discussed with Getty, for ease of reading.

# 1
# KIDNAPPINGS, HEARST AND HOUSES

He sat in an armchair, facing me, illuminated by the bright autumn sunshine. I would place a tape recorder on a table beside his chair, and he grew used to it. Sometimes he asked me to play it back to him. When the sun became too bright, he drew the middle curtains. While I sat in the light, he sat in the shadows, perhaps better to study my expression.

SDC: It's a lovely day. Have you been out in the garden?

JPG: Yes, I have.

SDC (*gesturing towards the sofa beside him*): What's the significance of the cushions with the *Ich dien* motto on them?

JPG: That's the Prince of Wales; they came from the room upstairs that he used to occupy.

SDC: I see. Now, do you feel in a position to talk today about the rather dramatic business of your grandson's kidnapping?

JPG (*a certain weariness appearing in his expression*): I don't know much about it. I have fourteen grandchildren, and they would all be at risk if I paid any ransom.

SDC: Yes, they would. I saw a small item in the *Daily Express* saying that an attorney representing the Getty family had offered £700,000 for the boy.

JPG: I don't think so; I don't think it's right.

SDC: Such a tragic story, and of course, it is a traumatic event in anybody's life. Naturally, we would want to refer to an episode of its kind and the stand you have taken in the matter. So, in principle, you're against surrendering to blackmail of any kind?

JPG: Yes.

SDC: Presumably, you would take a pretty tough line if you were the prime minister of a country whose planes were hijacked?

JPG: I think so, yes.

SDC: You find that this is a very dangerous policy to give way to?

JPG: Yes. It's blackmail. It may buy peace for a moment, but later on, it just encourages more demands.

SDC: There was a report in the *Daily Express* saying that you, in fact, were responsible for employing an ex-FBI agent, called Mr Chace, to find the kidnappers and track them down.

JPG: Well, I had nothing to do with that. Mr Chace works for the company; he doesn't work for me personally. Somebody had to deliver the money – it's not a very easy job to deal with the money – and he volunteered for it. The money, of course, was raised by the family, to which I contributed. It was a family matter.

SDC: The outcome seems to have been successful insofar as Mr Chace being able to contact them and then the police getting them.

JPG: I think the money was successfully delivered, but anything to do with the arrest can be attributed to the Rome police. They were responsible for that. I'm glad the boy was released safe and sound. [Getty's kidnapped grandson, J. Paul III, had his ear cut off!]

SDC: It must have been a very harassing time for you. It's certainly a most dramatic episode, but it's not an isolated one, is it? There's the case of William Randolph Hearst's granddaughter being kidnapped and shoved into the boot of a car. It was a dreadful story, wasn't it?

JPG: Terrible, yes.

SDC: You knew William Randolph Hearst, I remember. You must have been very shaken by the kidnapping.

JPG: Yes.

SDC: One wonders where it's going to start next. What

was Hearst like as a person? Did you have much to do with him?

JPG: He was a very interesting man, a very fine mind. He could write better, really, than any of his employees could.

SDC: He must have had certain interests in common with you because he was a great collector of works of art, and he took a great interest in architecture.

JPG (*chuckling*): I'm probably the only man who, as a guest at San Simeon, gave anything in return. I gave him some tickets for gasoline belonging to Tidewater Associated. I gave him, I think, 100 gallons [450 litres] of gasoline, or 500 gallons [2270 litres] – I've forgotten what it was – just as a memento.

SDC: Rather a nice memento, I think. Was San Simeon fascinating? Were you impressed?

JPG: Well, it's rather amusing because you could get yourself in a situation. I remember having the seat of honour right next to Marion Davies. She was on my left and Hearst was sitting across from me. Marion Davies liked to drink and Hearst didn't. He didn't drink very much, and he didn't approve of her drinking. Another thing he didn't approve of was taking food up to the bedroom, and she'd do that. He though it attracted rats and mice. Anyhow, she was sitting next to me and wanted another drink. Hearst said: 'No, you can't have another one.' I had a drink beside me that I hadn't touched, right by my plate, and a couple of minutes later, she reached over and asked me for my drink. Just as I was about to do so, Hearst boomed out: 'No, don't give it to her, she's had enough!' And she said: 'Do give it to me!' So, I was in the field of fire between the two batteries.

SDC: What did you do?

JPG: I didn't give it to her. It was Hearst's place.

SDC: He also wanted to have an interesting house in England and apparently left instructions with an attorney here that if either Leeds Castle in Kent or St Donat's Castle came on the market . . . 'surely buy'. Eventually, St Donat's came up, and he bought that.

JPG: He never spent more than a couple of days there.

SDC: He was very clever. He managed to get the people of

Boston Stump Church in Lincolnshire to part with the original medieval ceiling and had it sent down to St Donat's, because he said it had death-watch beetle and worms. Then, he raised a fund in Boston, Massachusetts, to preserve the decaying ceiling and with that money gave Boston Stump Church a new ceiling. Transferred down to Wales, the old one is a magnificent thing – powder-blue angels and gold.

He must have had a remarkable collection of things around him. Did it influence you in any way?

JPG: Hearst was here at Sutton Place once on a visit to Geordie, Duke of Sutherland. And, the tables I have in the dining-room came from St Donat's. They belonged to Hearst.

SDC: Well, you own a good deal of actual physical property which has gone up like everybody else's during the years . . . looking at it purely as a landowner. You or your company own this house, and it's an expensive house to maintain. You've got lovely gardens, quite a large estate here. You own a ranch house in Malibu, and you're building a fantastic museum.

JPG (*pointing to a framed colour photograph of the museum on a marble-topped table*): That's the place, there.

SDC: And then you've got a castle north of Rome. As a private individual your actual properties have been not only very attractive, but also very shrewd investments. Can you tell me a little more about these houses? For instance, the one north of Rome.

JPG: At the time, the company had a refinery in Gaeta, and I was interested in buying a house somewhere not too far away. So, I was taken by a friend of mine to see this house (*pointing to another framed photograph*) which was uninhabited and had not been inhabited, I think, for about a hundred years – the nineteenth century, anyhow.

SDC (*fetching the picture and indicating the more modern building*): What is this building here; what date would that be?

JPG: That's 1520.

SDC: Really, as early as that? And this appealed to you obviously as a task for restoration. [The castle beyond is of the fourteenth century. Getty rented it for seven years while he was restoring the sixteenth-century palazzo.]

JPG: Yes, it's very picturesque with the forest around it. So, I bought it . . . and it shows how you get publicity even when you want to avoid it. That was one real estate transaction that was headlined in all the Italian newspapers because the Odescalchis practically never sold anything. They'd owned the place for centuries. It made headlines in the Italian press also because it was sold to an American.

SDC: How long did it take you to restore your palazzo?

JPG: About three years.

SDC: Do you go there very much?

JPG: I haven't been there in two years.

SDC: Do you keep a staff looking after it?

JPG: I do, yes. I have friends there. Actually, I've had a convention there of all our ships' captains just in the last week.

SDC: And when you had it all done up, presumably you put a collection of furniture and pictures into it?

JPG: Yes, it was in ruins; there was no furniture.

SDC: So, have you got an important part of your collection in that castle?

JPG: No, the paintings are decorative-style, but there's some very fine sixteenth-century Italian furniture in the house, and some tapestries.

SDC: I think the thing that fascinates a lot of people in England is the fact that, with the whole world to choose from, you decided to make your home in England for a long period, the best part of fifteen years so far, and have lived the life of what one might describe as an English country gentleman. This is very flattering to the English, and you've chosen one of the loveliest houses in England: Sutton. What were your reasons for making that choice?

JPG: Although I picture the English country gentleman looking after the estate, in truth I'm looking after the oil business here. We have a superintendent to manage the estate.

SDC: Tell me again how you really came to live at Sutton.

JPG: Well, I had been thinking for years that I'd like to

buy a place. I was tired of living in hotels, and also, hotels have security problems. You know nothing about the employees. We were often in deep discussion about what we were going to bid at public sale or what we'd sell. We weren't the ones emptying the waste baskets. You don't know if your conversation is being recorded. Also, you're safer in a house like this than you would be in an office building, unless you managed the building – and that's a rather important factor.

Commencing, I'd say, in the summer of 1956, it became apparent to me that the Getty Oil Company should have an office somewhere in Europe from which to manage our eastern hemisphere operations. Other oil companies were doing this, and it seemed to me that Getty Oil should do likewise. The question was, where?

> **On Sutton Place**
>
> I had been thinking for years that I would like to buy a place. I was tired of living in hotels, and also, hotels have security problems. You know nothing about the employees. We were often in deep discussion about what we were going to bid at public sale or what we'd sell. We weren't the ones emptying the waste baskets. You don't know if your conversation is being recorded, whether the tapes are working or not.... When I came to England in May 1959 ... I was invited to dinner at Sutton Place by the Duke of Sutherland, and after dinner I was told by a mutual friend that Geordie was going to sell the place. Now, I didn't have any more idea of selecting Sutton Place as the Company's European headquarters than you have of buying the North Pole; but I happened to mention it to some English friends of mine, and they were very enthusiastic. They said it was such a wonderful place, so close to London.

I spent a lot of time in Rome, Naples, Milan, Paris, Geneva, Zurich, Brussels, and London in my regular travels, and I always kept the matter of an office somewhere in my mind. By the summer of 1959, I had come to the conclusion that the company must have such an office, and further, it was my opinion that, although I personally preferred Paris, in or around London was the place to have it. The majority of other oil companies had also selected London for their European headquarters. While some of the other cities had a lot going for them, when one considered tax laws, governmental attitudes, rent, and cost of operations, language difference, and so forth, it was obvious why London was the best choice. Therefore, in the summer of 1959, I began to actively look for suitable facilities. I was staying at the Ritz Hotel, and naturally, when my executives came to see me, they tended to stay at the Ritz. Our hotel bill was enormous! However, I had no intention of buying a house.

Shortly thereafter, I was invited to dinner at Sutton Place by the Duke of Sutherland, and during the course of my visit, I was told by a mutual friend that Geordie was going to sell the place. Now, I didn't have any more idea of selecting Sutton Place as the company's European headquarters than you have of buying the North Pole; but I happened to mention it to some English friends of mine, and they were very enthusiastic. They said it was such a wonderful place, so close to London, and that I'd never have another opportunity like this. The first thing I knew I was taking it more seriously. I must have visited Sutton half a dozen times, taking with me one company executive or friendly adviser after another.

> **On London**
> By far the biggest oil town in the eastern hemisphere is London. That is where most of the oil companies have their eastern hemisphere headquarters, most of the modern oil companies, and it is a great oil centre. The town is so big that you don't realize how much of an oil centre it is.

The Duke was quite anxious to sell – at least I felt he was asking an extremely reasonable price indeed. I just couldn't see how it could fail to be a good investment for the company, even if it should prove not to be feasible from a company headquarters operating standpoint. I think that subsequent events, thirteen years of efficient operation and the present price of land and houses, have proven me right. After the original purchase of the mansion and the land that went with it – about 280 acres [115 hectares] – the company acquired another 750 acres [300 hectares] approximately. Sutton is owned by a subsidiary of Getty Oil Company; it's not owned by me personally. It might be amusing for you to know that I pay rent to stay here like anybody else and that your lunch will be charged to me personally!

SDC: Nevertheless, you chose it partly because of your interest in domestic architecture, I take it. You were very influenced by the fact that it was built by Sir Richard Weston, wasn't it?

JPG: Yes, it was. Apparently, he was a good friend of Henry VII and Henry VIII.

SDC: There's an anecdote, isn't there, that he was walking round here with Henry VIII on the day that his son was beheaded in the Tower? [Francis Weston, the son of Sir Richard, was beheaded by Henry VIII for purportedly dallying with Anne Boleyn, a frequent visitor at Sutton.]

JPG: Yes.

SDC: After a good lunch, no doubt! Sounds a little cynical. I didn't realize he'd been a friend of Henry VII, too. This house was built in Henry VIII's reign, wasn't it?

JPG: Yes, in the early days.

SDC: What, about 1520?

JPG: 1521.

SDC: It certainly is one of the most remarkable houses in England. You've been able to conduct all your operations against a very lovely background.

JPG: It's worked out, and the property's advanced in value very much.

SDC: I'm sure it has. At that time, Geordie thought it was rather a white elephant. He wasn't asking a great deal for the house. I think the quoting price was £40,000–50,000 for the house with 60 acres [25 hectares], including the drive and those interesting lodge gates. Did you take over much of the contents from Geordie or is it all yours?

JPG: I bought it furnished.

SDC: Do you like living in England, on the whole?

JPG: Yes, I do; however, I'm not crazy about the weather

here. I didn't leave California for England because I thought the weather was better!

SDC: I'm anxious to ask you one question about the business of making money. Obviously, if your object was to make enough money to buy all your works of art and live the good life, you could have stopped years ago, or you need never have carried on after you retired once as a millionaire at twenty-three.

JPG: Yes.

SDC: And then you lived the good life for a short time and got bored with that?

JPG: Yes.

SDC: What was the drive then – a desire to control a big organization?

JPG: I wouldn't think that! I suppose I was in a rut!

SDC: You must be a perfectionist in business, really. You always try to prune out the bad staff and get the better staff. I was rather amused by the poster I saw in the secretary's office: 'Anybody found dead in an upright position will be dropped from the payroll.' Did you hang that up?

JPG: No, the girls did.

SDC: Because you're driving them too hard?

JPG: I don't know what their motive was.

SDC: They really don't think that, do they? They like you very much; you seem to be a very popular employer, and this comes through from your earlier experiences in oil rigs where you always worked with people.

JPG: Yes.

SDC: And you really enjoyed your life at work, I suppose?

JPG: Yes. I never had any explanation of the business success, so-called. I've just tried to look after my business and be competitive. I like to think that Getty Oil is a very competitive company.

SDC: Business is the breath of life to you, isn't it? To conduct the operation?

JPG: I've always thought that I had a great gift for being a beachcomber, being an idler. I don't want to just sit around and look at the sky, but I could do very well with a beachcomber's life – walk on the beach, swim, ride a horse, read books, listen to the waves. But I've always had work to do, and if I could find someone who could do the work better than I could – and I've always said this – I would hire that man and take it easy myself! I don't work just because I need the work. I'm not unhappy if I'm not working; and I've never been personally motivated in my work to where I was worried about other people being better than I was. In a few instances, I've found men who could do parts of my business as well or better than I could myself; and I've always been happy at finding someone who could take a load off my shoulders. But as they say, the old blind sow brings back an acorn occasionally!

SDC: He brings back quite a few. You haven't found this man you feel you could hand it to, at all, yet?

JPG: Well, I like to think I am impartial. Sometimes you get enthusiastic about somebody, and then he does something that you feel was a real boner. For example, several men were praised to the skies for years as leading business executives and are now in eclipse. One has been particularly mixed up in politics. There's an old saying that the head of a company is just as good as the stock. Mr X has one of the largest companies in America and has been an outstanding success; he's built up and increased his assets many times over. Yet today, the company's under a cloud. The shares are weak, and he's been strongly criticized. The press is after him. And this man was considered the number one business executive in America! Yet I think he should be classified as a professional manager.

It's like the story Colonel Stewart told me. Colonel Stewart was the president of Standard Oil Company of Indiana and was one of Rockefeller's top men. He probably was number one in Rockefeller's estimation. However, he got into the Teapot Dome affair – was heavily involved. Rockefeller sent for him and said: 'Colonel Stewart, I regret very much that I must ask for your resignation.' Stewart said to him: 'Well, sir, you always said I was your best executive. What's caused you to say what you've just said?' And Rockefeller replied:

'Colonel Stewart, I agree that I said you were my best executive, and you always have been. You've built the Standard in Indiana to a tremendous degree, and I've been highly satisfied with you. But you have involved me in a quarrel with the United States Senate, and I have large business interests in the United States. I cannot afford to have my business interests in the United States involved in your quarrel with the Senate. So, I must ask for your resignation.' Stewart told me: 'I thought the old gentleman was wrong, but I can see now that he was probably right.'

SDC: How interesting. Have you managed to keep clear of any problems of that kind yourself?

JPG: Well, I think so. You never know when the lightning may strike. There are men who are extremely successful but lack judgement. Some of the men who have the best judgement have no drive, so they never get anywhere. And some of the men who have tremendous drive and are highly intelligent don't have good judgement.

> **On Napoleon and Hitler**
>
> I think that, had I been living at the time and had been a close friend of Napoleon, I would have said to him, before he advanced into Russia, that he shouldn't do it. That he should try to conciliate people rather than antagonize them. Be less keen on fighting. He had too much to lose. I think if he had nothing to lose, he might be looking for strife and revolution. But if you've got a lot to lose, why would you? For example, I can't understand the mentality of Hitler. Hitler had a tremendous position in Germany; he was the Dictator of the German people; he'd been successful. The German people were supporting him, and he had tremendous power and prestige. I think he should have started a policy of conciliation: conciliate foreign countries, conciliate the Jews, continue building his good roads, opera house and such – spreading music. 'The pitcher that goes to the well too often gets broken.'

SDC: You once mentioned that Napoleon had great drive but lacked judgement in going into Russia and also Spain.

JPG: I think that, had I been living at the time and had been a close friend of Napoleon, I would have said to him, before he advanced into Russia, that he shouldn't do it. That he should try to conciliate people rather than antagonize them. Be less keen on fighting. He had too much to lose. I think if he had nothing to lose, he might be looking for strife and revolution. But if you've got a lot to lose, why would you? For example, I can't understand the mentality of Hitler. Hitler had a tremendous position in Germany; he was the Dictator of the German people; he'd been successful. The German people were supporting him, and he had tremendous power and prestige. I think he should have started a policy of conciliation: conciliate foreign countries, conciliate the Jews, continue building his good roads, opera house, and such – spreading music.

SDC: But he wouldn't have been Adolf Hitler if he had had that kind of balanced judgement. It wasn't in his character, was it?

JPG: 'The pitcher that goes to the well too often gets broken.' Napoleon was successful in a hundred battles. Hitler was successful in many of his major moves, but they always wanted more.

SDC: And he wouldn't listen to people. Napoleon didn't want to know the facts, didn't want the truth. You've never been like that, have you? You've always got the facts.

JPG: I try to run my company more conservatively than a professional manager would run it because a professional manager is living on a salary; he probably doesn't have any stock in the company, and if the company goes broke or gets into difficulties, his personal fortune is not jeopardized because it's not invested in his company. So he might take a risk that I wouldn't take in my company. I like to be very cautious and conservative.

SDC: You're cautious and conservative, but you do gamble, don't you?

JPG: Yes.

SDC: At the end of it all, when you've got all the facts, you do gamble.

JPG: But I have to have the facts. As I say, if I wanted to gamble on typical gambling games, I'd buy a casino and have the percentages for me, rather than play. Augustus was cautious.

SDC: Yes, he was. And successful for that reason. You're rather an Augustus Caesar in your temperament, aren't you?

JPG: Yes.

> **On gambling**
> I have to have the facts. As I say, if I wanted to gamble on typical gambling games, I'd buy a casino and have the percentages for me, rather than play.

SDC: Because you have always weighed the consequences very carefully before making a bold decision.

JPG: Yes.

SDC: Would you say that Augustus Caesar is one of your heroes?

JPG: I think his philosophy was closer to mine than some of the others.

SDC: More so than that of Julius Caesar?

JPG: Yes, more so than Julius Caesar, although I must admire Julius Caesar more than Augustus because he was probably an abler man. But if you take five very able men as shown by their success – Julius Caesar, Augustus, Napoleon, Hitler, and Mussolini – you'll find that the only one who was really cautious was Augustus. The rest of them seemed to think that their success would go on forever. Mussolini got away with a lot of daring deeds. He took a lot of chances when establishing himself. However, it was a terrible mistake to make when he got Italy into the war because he had been popular with Roosevelt, with Churchill, and with Stalin. If he'd kept out of the war, he'd have been popular with Hitler, too. He had everything to gain; he would have got rich if he'd kept neutral because he would have pleased everybody.

# 2
# BUSINESSES AND BUSINESSMEN

My car finally turned a corner beside well-known grass and woods, and the great Tudor house with its open courtyard came into view. I drew up in front of the main entrance. The butler, Bullimore, by now a familiar figure, greeted me warmly, took my hat and coat, and led me through to the drawing-room. Paul joined me shortly thereafter and, having checked the tape recorder, he began by asking the topic for the day. I suggested the business world and all its complexities.

SDC: I suppose, in the end, you were really destined to be a businessman. I was reading your account of your return from Europe in 1914 and the early days of your business with your father, George F. Getty Inc. You began to notice ways in which you could trim the expenses and increase the output of the company. Did your father react kindly to this sort of investigation?

JPG: Well, he was ill at the time, and he got one viewpoint from me and one viewpoint from his staff, which was opposed to mine. He wasn't in a position healthwise to ascertain the facts.

SDC: You've approached business affairs with a pretty ruthless eye throughout your life, haven't you? I've noticed that again and again you go through the accounts of a company and decide where it is wasting money, where it can cut down.

JPG: I think I can call a spade a spade.

SDC: There is a most entertaining anecdote of yours in the book *How To Be Rich* [Playboy Press, Chicago, 1965], where you decided on one occasion to propose something to a meeting of directors which you deliberately made ridiculous, one that would harm the company's affairs, in order to see which of them would spot the flaws and which would just say 'Yes, sir,' and agree.

JPG: That's true. As it turned out, only one man was trying to protect the company; the others were following on with me, either unthinkingly or with knowledge. I was reminded of the motion picture producer who said: 'For heaven's sake, don't say "Yes" before I do!'

SDC: But you really felt that these sycophants weren't really contributing?

JPG: Or else they hadn't analysed the proposition. I haven't encountered it as much in the companies I am in nowadays. Those people have been weeded out long ago.

SDC: Were they people you had chosen yourself or were they inherited from some other system?

JPG: They were inherited.

SDC: So, you were watching to see how they would react. What do you do if you find a company is going into the red for some reason?

JPG: If it's headed for the rocks, I change the course.

SDC: Have you often had this problem where companies you have taken over seem to be heading for the rocks?

JPG: Yes. I'm not mentioning names, but once, years ago, I took over a company and the man who had owned it said he wanted $50,000 a year for himself. However, he thought $200 a month was too much for anybody else. It was noted that he had about six managers there in the last three years, and any manager who asked for an increase in pay was automatically fired. So, many of his competitors had good managers, former employees of this gentleman. Some of them were outstanding executives, but they were ruthlessly fired the moment they suggested a raise.

SDC: I noticed, on one occasion, you docked $5 a month off the pay of some of your leading executives to see whether they would notice. They soon noticed when it came off their own pay?

JPG: Yes. It's a strange thing, for example, that when you lend your car to a friend, and he gets a couple of scratches on it, it doesn't seem to bother him much. But if you borrow his car, and you get a couple of scratches on it . . .

SDC: You find that very much in life. You're a strong character, obviously. You don't mind having a discussion when a man comes along and says: 'What's this about my $5?' What do you say to him?

JPG: Well, I would say that I didn't think it was good business. As a matter of fact, I sent off a telex this morning restricting any commitments by my museum over $10,000 without my prior approval. The reason is a bill that had been approved by my staff for $62,000, for signs.

SDC: Just pointing the way around the room?

JPG: Apparently so, yes. Either they've got 62,000 signs there, or . . .

SDC: I think they cost more than a dollar a sign now in Los Angeles.

JPG: But even at $10 a sign . . . that would be 6200 signs! As a matter of fact, I would like to say a few words about wealth.

SDC: It would be very interesting to hear your views on that.

> **On life's inequalities**
> There's always the best hotel in town and the best room in the best hotel in town, and there's always somebody in it. And there's always the worst hotel, the worst room in the worst hotel, and there's always somebody in that room, too.

JPG: Talents and rewards are not equally distributed in any field. For example, politics. I've known men who were very intelligent, very hard-working, very honest, very able in every way, good speakers; and yet they never got beyond the city council. Yet, another man becomes governor of a state or president of the United States or prime minister. What is the explanation? Why does one man who seems qualified to be in a high political office never make the grade? We find that also in the entertainment field; we find some people who seem to have all the qualities of a movie star. I've known girls, myself, who had looks, personality, were photogenic, hard-working, were good actresses, and yet they never got beyond an 'extra' part. Then, some girl who wasn't so good-looking, didn't seem to be such a good actress, becomes a great star. Then, we have the field of painting; some of the best painters couldn't sell their paintings. They died bankrupt.

SDC: Like Vermeer and Rembrandt, who died in poverty.

JPG: Yes, and Van Gogh. And so it is in business. Whether it's an Iron Curtain country or in the West, there's always the best hotel in town and the best room in the best hotel in town, and there's always somebody in it. And there's the worst hotel, the worst room in the worst hotel, and there's always somebody in that room, too.

> **On life's inequalities**
> 
> I've known girls, myself, who had looks, personality, were photogenic, hard-working, were good actresses, and yet they never got beyond an 'extra' part. Then, some girl who wasn't so good-looking, didn't seem to be such a good actress, becomes a great star.

SDC: Yes, you're dead right.

JPG: It might seem disproportionate that some people make a great deal of money in business and some people who seem equally able, equally hard-working, equally competent, never make more than a small amount of money and haven't very much success. But, that seems to be the way the world goes.

SDC: I can understand your analogy with the entertainment world, but when you compare making a success in business and making a lot of money with the fate of people like Rembrandt who died in poverty because they were not sufficiently recognized at the time as artists, surely there is a difference. The yardstick of success in business is financial success, on the whole. Of course, it also includes the ability to organize and run an efficient business, but when you're talking about a person who accumulates a vast fortune by his own efforts, there is a visible standard by which you can measure that. It's not a question of opinion, like an expression of art.

JPG: Yes. Look at the case of the Coca-Cola Company. The man who started Coca-Cola didn't live long enough to see it develop into a great company.

SDC: So he may have died in comparative poverty.

JPG: Yes. And so that could be true of other businesses. I'm thinking, too, of the question of responsibility, what the consequences would be if people were levelled out. I would think that if you divided all the money and property in the world at three o'clock this afternoon, half an hour afterwards there would be a lot of people who had nothing. Those people had given theirs away, lost it, gambled it away, or for some reason or another just didn't have it anymore. And, you'd have practically as great a discrepancy in wealth in half an hour as you have today. It may be some exaggeration, but not much.

SDC: And from this, you are drawing a general philosophy about capitalism?

JPG: Well, it is true that it can very easily bring about the disappearance of the rich. The rich are few. I suppose it could bring about the disappearance of political figures. I think nature works on that principle, and it's hard to change it.

SDC: I agree.

JPG: Why should some people live a long and happy life and others die very young after great suffering?

SDC: So it is a basic factor of nature?

JPG: I think nature doesn't create people equally. Certainly, we don't all have a singing voice like Caruso, do we? Or Nellie Melba? I think that businessmen who have made a fortune generally have made it honestly. It's hard to make money in business. It's very difficult. You've got to serve the public, and if your reward is relatively great, that's the way nature works. I don't see that it's any more extraordinary for a man in business to accumulate a large fortune than it is for a man in politics to attain a very high office.

SDC: I think that's a very good analogy. We can't all be prime ministers.

JPG: No. A prime minister may not have as much money as a successful businessman, but he has a great deal more power and a great deal more position. But why should one man be prime minister and another man be on the county council? Why should one man be a millionaire and another man be a very modest businessman?

> **On business**
> It's hard to make money in business. It's very difficult. You've got to serve the public, and if your reward is relatively great, that's the way nature works. I don't see it's any more extraordinary for a man in business to accumulate a large fortune than it is for a man in politics to attain a very high office.

SDC: But you're appealing to a slightly different electorate, aren't you?

JPG: Yes, but you have to perform a service to the public; and in a free country you have to face competition, which is remorseless.

SDC: Let's talk about your father for a moment, if we may. He was a very prudent and thrifty man who didn't believe in borrowing money at all if he could avoid it.

JPG: True.

SDC: Therefore, I don't suppose he ever could really expand the business in the way you have done. From an historical point of view, if one is looking for a parallel, one would naturally think of Philip of Macedon and Alexander the Great, where the father had established the position of Macedon, and it was left to the son to build on that base, spread it out all over the near-Orient and create the fantastic empire he did. Were you conscious when you took over the reigns after your father died that you were going to end up with one of the world's greatest businesses? Or did it just expand, just grow?

JPG: I never thought about it that way. I thought that it was a good size business, and I wanted to be competitive; I wanted to build it up within reason. I never suffered from megalomania. One thing I can say about our business is that we have grown through internal growth; we haven't merged our way into bigness. Many companies grow big by merging with other companies. We've never merged with an outside company. Now, it's obvious that I would double the size of the company by merging with an outside company of the same size. Then, the company that survived would be twice as big, and it's rather easy to grow that way. But we took it the hard way ... by internal growth. It's true we merged Tidewater and Getty Oil in 1967, but we'd already bought about 70 or 80 per cent of the Tidewater shares from our own internal cash resources. We haven't borrowed money in order to expand Getty Oil.

SDC: Do you feel that you've just established a kind of built-in system of growth through efficiency because it has spread and spread and got so big that it's almost as if it has grown by a law of nature, but it isn't so?

JPG (*laughing*): No.

SDC: Obviously, by a law of Paul Getty! But one wonders why it is that you have been able to build up such an enormous personal company when, as you say, most other companies owe their existence to mergers, to the increase, no doubt, of public shareholdings in their affairs. You've kept pretty much in control of the actual shares yourself, haven't you?

JPG: Yes, I have about two-thirds of the shares.

SDC: And who owns the other third?

JPG: The public ... 30,000 people.

SDC: So, you're responsible to 30,000 shareholders. However, it must be the biggest single company in the world owned to the extent of two-thirds by the people who started it?

JPG: I suppose it is. I think Krupp is changed. It used to be owned by one man.

SDC: And yet you've managed to retain control of all two-thirds.

JPG (*grinning*): One of the last of the Mohicans.

SDC: Do you make all of the major decisions now?

JPG (*with a ghost of a smile*): Well, I like to think that I'm consulted about them. But, of course, a company – technically speaking – is run by its board of directors. Actually, the directors have the final say.

SDC: Do you attend their meetings?

JPG: The board has never met in Europe. It meets in America.

SDC: What is your exact position in the company?

JPG: I'm President.

> **On business**
> *What is your exact position in the Company?*
> I am President, and Chief Executive Officer. We have no Chairman. I own 62 per cent of the shares.

SDC: President, obviously.

JPG: And Chief Executive Officer.

SDC: So, you really do run the company.

JPG: Yes, we have no Chairman.

SDC: I see. Do you find it more exciting to run the big business it is than when it was a smaller business?

JPG: I think it's less exacting because in a big business you've got so many resources. You've got company attorneys, geologists, accountants; you've got so many people to help you. Whereas, when you're in a small business, you're in by yourself. When I started my first wells, I felt that I was completely responsible. I was President, Vice President, Secretary, Treasurer, General Manager, Assistant General Manager, Chief Tool-Pusher, Assistant Tool-Pusher, Purchasing Agent, Assistant Purchasing Agent. When I was drilling for myself, I was the whole office works. Actually, it was very much the case with Getty Oil when it started.

There's a lot of difference between being in a big business and being in a small one. In a small business, a man knows how much money he's got, whether he's making or losing money. He has a payroll to meet next Saturday, and he knows whether or not he's going to make that payroll. He knows how much money he's going to take in, and he could tell you right now what his balance is. Whereas, in a big company, you've made a loss, maybe in various departments, but until the fifteenth or twentieth of the succeeding month, nobody except the controller would know the amount of the bank balance. It's about on the same basis as the Post Office is run. I suppose very few people can tell you, except those in charge, just how much money the Post Office has in the bank right now. And, I suppose very few people could tell you, if you go down to the Post Office to mail a package to Los Angeles weighing 8 ounces [230 grams] and you put a stamp on it, whether the Post Office is going to make or lose money on it. Most of the employees at the Post Office couldn't care less.

SDC: Looking back on it all, does the tussle with Tidewater Oil stick in your memory as the major tussle of your career?

JPG: I think so, yes.

SDC: Because they were very unsporting, weren't they? The board of Tidewater fought you very hard to stop you getting control.

JPG: Yes, it was equivalent, of course, to major companies taking over small ones. There are very few instances of a small company taking over a major company. It's like they say, the Hilton might take over a smaller hotel, but it's not likely that the Talbot in Ripley would take over the Savoy.

SDC: So, you feel that's what you were really doing?

> **On business**
> You never know how powerful, in fact, an outfit like Standard Oil of New Jersey is until you get in a fight with them. I don't suppose I could have borrowed a dime from any bank in the United States.

JPG: In effect, yes. You never know how powerful, in fact, an outfit like Standard Oil of New Jersey is until you get in a fight with them. I don't suppose I could have borrowed a dime from any bank in the United States.

SDC: They were really trying to seal off your credit?

JPG: Yes.

SDC: But you managed, with the help of your mother, to get about $4½ million for this particular contest. At what stage did you really make the breakthrough that enabled you to be the big operator and stop the others?

JPG: It was New Year's Day, 1935, at San Simeon. Jay Hopkins, who was the founder of General Dynamics Company and a great friend of mine, called me up and told me that John D. Rockefeller Jr would sell his rights, his shares in Mission Corporation, which had been organized New Year's Eve, and he wanted to know if I'd be interested. I said of course I was! So, I bought them over the telephone.

Farrish and Teagle, who were the heads of Standard Oil in New Jersey, had been trying to locate John D. and tell him not to sell. But they couldn't locate him because he apparently was travelling on a train on his way to Arizona. Anyway, to make a long story short, when Rockefeller returned a few days later, they said: 'Oh, Mr Rockefeller, we were very anxious to get in touch with you. We wanted to tell you to be sure not to sell your stock in Mission Corporation because we're in a big proxy fight. So he said: 'Why, I'm sorry, I didn't know. I've already sold my shares.' And they said: 'You have? Who did you sell them to?' And he said: 'Well, I really don't know the man at all. I understand he's a very nice young man, but I can't remember his name.' And they said: 'His name wasn't Getty, was it?' He replied: 'Yes, I think it was.' They answered; 'Oh, my God. That's the man we're fighting!'

You have to have the opportunity. If you don't have the opportunity, you can't cash in on it. I recognized the opportunity instantly. It didn't take me a second to make up my mind. Of course, I can't take any credit for influencing the Jersey company to organize Mission Corporation or for influencing John D. Rockefeller to offer his rights for sale.

SDC: But having got it, what was the reaction? We know that the Tidewater people were furious. What did they do about it?

JPG: Well, they were naturally disgruntled. They held on

as long as they could. We had another proxy campaign, as explained in my book.

SDC: Was there a lot of personal animosity at the time?

JPG: I don't think there was. I've always tried to separate business from pleasure, and I've not been angry at men because they've differed from me in business. I've been good friends with people who were adversaries, competitors of mine in business.

SDC: And they accepted that, too?

JPG: Yes. In fact, I have had a high opinion of most of the people I've been in competition with in business, found them very able. And my father was that way, too. He didn't carry his business into his social life. I wouldn't see any reason, for example, if we were competing in an election, and you were a candidate and I was a candidate, why I should be angry at you unless you did something outrageous. But the mere fact that you were polling for votes and I was polling for votes, even if you were more successful than me, I don't see any reason why I should be angry with you.

I had a visitor here yesterday afternoon, a famous man in American business. He's head of a big conglomerate, and he was interested in making a deal with me in connection with a business I have. I was rather amused because he said that people in the States think he is crazy when he said he was going over to make a deal with Paul Getty.

SDC: Why? Did they assume automatically that he'd get the worst of the deal?

JPG: He said that he was not likely to get anything intelligent when he said he thought he was going to make a deal with me. I told him the story of Harry Sinclair, who was riding up from Jersey City to Fifth Avenue in New York about 1932 with Elisha Walker, who was a great financial man at that time (Kuhn, Loeb, and Company). So, Harry said to Elisha: 'I just made a deal with Paul Getty.' And Elisha said: 'Well, I'll tell you one thing, Harry.' Harry said: 'What's that?' Elisha answered: 'I'll bet it was a devil of a swell deal for Paul Getty!' Harry was rather put out by Elisha's remark because he thought he was a pretty good dealer himself; but I have made one or two deals in my life that were not bad!

SDC: But the other people who make deals with you hope to get something out of it. As they say, we must let the other man make a profit, too.

JPG: Well, yes. I always like to see the other man make a profit. My father sold an interest in a lease in Oklahoma to a man, and I said to him after the man left: 'Don't you think you could have got more for it?' My father said: 'Yes, I think I could have, but you must never try to make all the money that's in a deal. Let the other fellow make some money too, because if you have a reputation of always making all the money there is in a deal, you won't make many deals.'

SDC: I can see that entirely. Did the man who came to lunch with you yesterday from America succeed in doing a deal?

JPG: Well, we're still discussing it.

SDC: In the Second World War, you were running an aircraft factory, weren't you?

> **On business**
> My father said: 'You must never try to make all the money that's in a deal. Let the other fellow make some money too, because if you have a reputation for always making all the money there is in a deal, you won't make many deals.'

JPG: Yes, I was.

SDC: You had your oil companies going as well. Who looked after them?

JPG: The regular management.

SDC: And you had no problems with the company getting into difficulties?

JPG: No. They kept regular hours and had their regular work, whereas I was working double the hours I formerly worked: sixteen or seventeen hours a day, seven days a week.

SDC: The aircraft industry is very changeable, isn't it? You were not, presumably, ever manufacturing jet engines when you were in the Spartan Aircraft Company?

JPG: No, we didn't manufacture engines, we just manufactured planes. Of course, when you manufacture in the free enterprise system, you appreciate the advantages of having a factory behind the Iron Curtain, where you change models when you want to change. This side of the Iron Curtain, you have to change just about the time that you really learn how to do the job. You're making an engine and you learn how to do it, then after several thousand engines have been manufactured, you really learn how to do the job. Then, of course, the competition has got something that is improved, and you have to have an entirely different engine. So, you go back to the learning curve. People learn by trial and error in the manufacturing business, and finally, after thousands have been produced and you've served a long, hard apprenticeship, you know how to manufacture them and you do an excellent job; and then you have to change again! You never get a long run at anything. Someone comes along with a better engine, a better car, a better refrigerator, and you have to meet the competition; you have to redesign the whole thing. It's a constant learning curve.

SDC: What are some of the other things you remember?

JPG: Well, there were thousands of people to train, and as soon as you got a man trained, he was drafted. You had to replace him with somebody with no experience. It really was a tremendous job, and of course, my background in aircraft manufacturing was zero minus zero, so I didn't have much to go by. (*Laughing*) I often thought that I'd have made a good assistant to someone who really knew. But even so, it's surprising what you can do if you really apply yourself, and we did pretty well. I was one of the few heads of manufacturers who didn't get into financial trouble. Most of them took fixed price contracts and lost their shirts and their companies.

SDC: Yes, like Rolls-Royce with the RB–211.

JPG: Yes. I had a hunch and was fortunate to avoid that type of contract. For instance, a friend of mine got a contract for a gun mount, 2500 gun mounts. I suppose it was about $1000 each. After the first hundred, you begin to know what your ultimate costs will be; and he could see that at the end of his contract the gun mounts would be costing him about $4000 each and the early part of the contract would be costing him much more, so he was bankrupt. He had to go to the prime contract and say: 'Here's the company.'

SDC: You managed very shrewdly to avoid the pitfalls that beset many contractors in arms.

JPG: Well, if you're pretty conservative by temperament, you're apt to give a more realistic estimate than most people would. I built up a reputation for being factual with my views, being realistic. It's easy to say that you're going to redecorate a house and you're going to do it in so many days; but actually, it generally takes longer than you expect, doesn't it?

SDC: You didn't retain your interest in the aircraft industry after the war, did you?

JPG: I sold it not long ago. Just a few years ago. We started making caravans instead of airplanes. We made 25,000 large trailers.

SDC: And then you sold the company altogether?

JPG: Yes, then I sold the company.

SDC: You don't seem to have stuck to oil by any means as an exclusive preoccupation. You went into the aircraft business, and you also invested a great deal of money in other companies.

JPG: Yes, it's been my misfortune to always be going to school. I had the grounding in the oil business, producing oil. I knew the business pretty well and obviously, when you know something well, you can do it. It's just like a cook. If he's a good cook, he knows the kitchen thoroughly and knows how to cook everything under the sun. Whereas if you and I went into the kitchen and had to cook a meal, it would be chaotic. We wouldn't know the kitchen, the layout, how to light the stove, and we wouldn't know how long to cook the meat. We'd probably burn it or have it only partly cooked. Some things we're supposed to cook, we wouldn't have a clue as to how we're supposed to do it, and we'd be very tired. I had this airplane business, which was very tiring for me and then after the war, we started making trailers. I'd had no background in that, and it was tiring. I had to go to school and learn about it.

SDC: Go to school all over again?

JPG: Yes. Then, we got into so many other businesses that were strange to me. We got into financing trailers, which you call caravans. We'd finance them and insure them, so we had our own insurance company and our own financial company, so I had to learn those businesses. It wasn't easy because I'm not bright enough to learn a business in ten minutes. It takes me time.

Then, we got into foreign oil exploration, went into the Middle East where we got a concession and were fortunate enough to get production going. Then we found we had to go to school because we were doing all sorts of things you do in an underdeveloped country that you never think of doing in a developed country. For instance, we found ourselves building roads and cities.

SDC: Going back to the war for a moment, you referred in one of your books to the fact that you rather wanted to get into the First World War when America joined but weren't able to do so. What happened then exactly; what were you doing at the time?

JPG: Well, everybody was subject to the draft in my age bracket, but they chose the numbers in a sort of wheel lottery; and my number didn't come up. I enlisted in the Air Force and nothing happened. Then I enlisted in the artillery. I had a cousin who was in the heavy artillery, and he liked it very much. It was rather amusing the way he described it. He said that a lot of people avoided the heavy artillery because they thought it was heavy lifting, but actually, everything was done by machinery. They took very good care of the heavy guns because the General would rather lose a regiment than lose one of those 16-inch rifles.

SDC: You never did, in fact, get called up?

JPG: No. I had volunteered as I say for the heavy artillery about eight months before the war ended. I passed the physical examination but never was called up. I got a letter from the Secretary of War commending me on my patriotism in volunteering and so forth. I showed it to my cousin, who had about two years of actual fighting, and he didn't get a letter from the War Department, so he felt wronged. He was very loud in his criticism of that state of affairs! (*Getty laughed heartily at the recollection, his shoulders shaking with amusement.*)

SDC: What did you do during the war?

JPG: I continued in the oil business.

SDC: So, your most consistent love has been business?

> **On business**
>
> I would regard business as something like being thrown off a bridge into the water, and you want to keep afloat, so you start swimming. You swim until you get out of the water.

JPG: I wouldn't say it's been a love of mine. I would regard business as something like being thrown off a bridge into the water. You find yourself in the water, and you want to keep afloat, so you start swimming. You swim until you get out of the water. So, you get into business. I didn't have any great love of business; I'm not the sort of man who would be nervous and upset if I didn't have any business. I want to do something. I don't want to just sit around and look at the ceiling for a whole day.

As long as I am in business, I will give it sufficient attention and time and effort to discharge my responsibilities. Now, I don't particularly like parrots . . . But if you had a collection of parrots, say fifty parrots, and you asked me as a favour – a great favour – to you to look after those parrots while you were gone, and I accepted the obligation, I think I would have a feeling of being conscientious. I would look after the parrots, even though I didn't like them. I'd see that they were properly fed, had plenty of water, and their cages were kept clean. I wouldn't want to let them starve to death for lack of attention or grow thirsty.

SDC: Why is it that a man who has been so successful and made so much money and has reached a position where he could at any moment sell stock or hand it over to somebody else and retire with a very comfortable fortune with a very peaceful existence in the countryside not do it? You could read books, buy works of art. Yet, you have continued to retain iron control of a major industrial empire. Have you ever been tempted to jettison your active control and responsibilities in favour of cashing in and living on the proceeds?

JPG: I've thought about it, of course. It's one of the problems of having the reputation of a rich man. The only way I could be rich, in the way of having a lot of surplus cash, would be if I just sold out and reinvested the money. Naturally, you find people who are much bigger in business than I am – Shell Company and BP, Standard Oil of New Jersey, the Exxon Company now – they borrow hundreds of millions of dollars, and they don't just do that to prove they can borrow it. They need the money. And I don't think you feel particularly rich when you're borrowing money. You find you don't have enough funds, so you take care of your obligations and have to borrow from the banks. There's no point in being rich at that stage.

> **On business**
> As long as I am in business, I will give it sufficient attention and time and effort to discharge my responsibilities. Now, I don't particularly like parrots. I don't have anything against parrots, I just don't particularly like them. But if you had a collection of parrots, say fifty parrots, and you asked me as a favour – a great favour – to you to look after those parrots while you were gone, and I accepted the obligation, I think I would have a feeling of being conscientious. I would look after the parrots, even though I didn't like them. I would see that they were properly fed, had plenty of water, and their cages were kept clean. I wouldn't want to let them starve to death for lack of attention or grow thirsty.

SDC: But this is while you're conducting the operations of a large enterprise and exploring new fields in Alaska or the North Sea. We can see that, as a controller of a big business, you might need to borrow large sums to lay out on these projects. But that doesn't alter the fact that at any given moment, presumably, you could sell the thing for £1000 million, £2000 million or whatever it is and literally put that into shares, blue chip shares, and live on the proceeds. But you've never felt you wanted to give up control.

JPG: I've never come to that conclusion. I've thought of it two or three times, but there are problems. In the first place, if you get a lot of money, what are you going to do with it? Unless you give most of it away, you've got to reinvest it. And also today you pay a big capital gains tax, a third or so of your fortune. And you reinvest in things that are probably not any safer than what you're in now. And you may have less responsibility, you know, in shares in a hundred different companies. And if you put your money into bonds, then you've got inflation which takes away from the value. Bonds are for people who have dollar obligations, like banks, insurance companies. They are not for private investors who want to keep the purchasing power of their investments and have their investments increase in value, increase in purchasing power.

SDC: In fact, you haven't devoted a great deal of your time to idleness, have you? You work very hard.

JPG: It's unfortunate that I was driven to it. I've never worked just for the sake of work. I'm not one of those people who think 'Oh, my goodness, what am I going to find to do today?' I'm generally swamped with work; and I am, if I say so myself, conscientious. I work even though I don't like to work because I feel it's my duty to work. As long as I'm active in business, I have an avalanche of things to do. I wish I were more intelligent, that I could work faster. I wish I could do fourteen hours' work in five hours, but I've never been able to do it.

SDC: This has often been regarded as a mark of a great man, or great statesman, this ability to give unwearied attention to detail for many hours on end. Napoleon did this, and you've always been able to do it, although you may have been going against the grain. You have, in fact, devoted the necessary time to get through your work.

JPG: I always remember what my father said: 'No man's judgement is any better than his information.' It's easy to act on 40 per cent of the information, but it's harder and it's better to act on 90 per cent of the information. You'll probably never have 100 per cent of the information. All you can do is the best you can do. But it takes longer to fill in the picture will all its details and not just take a quick glance at it. But I would say this. I've never come to grips with success in business and what it's due to. As I said earlier, at my stage in life, I can look back over fifty or sixty years, and I can see the men of my own generation who have impressed me, whom I thought were the ones who would be successful, but it hasn't worked out that way for them. Some of those I thought were rather sloppy have made successes, but actually very few do. If anybody think's I'm successful, I don't have any particular reasons to explain it. I don't think I've worked harder; I don't think I've worked longer hours; I don't think that essentially I've been more intelligent than many other people have been who have been less successful. So, I can't tell you just what you need to be successful in business.

---

**On business**

I always remember what my father said: 'No man's judgement is any better than his information.' It's easy to act on 40 per cent of the information, but it's harder and it's better to act on 90 per cent of the information. You'll probably never have 100 per cent of the information. All you can do is the best you can do.

---

SDC: Ceaseless application seems to have been a large part of the recipe. You seem to have devoted a tremendous lot of detail, attention to the facts.

JPG: And so have many other people.

SDC: And then on top of that, do you think a willingness, when you've made all the calculations, to gamble a little more than some people?

JPG: I think that's part of it. I said in my book why some worship at the shrine of security, yet some more than others.

SDC: And you seem to be able to devote an immense amount of energy to work still. Do you find that you can still get through as much work as you used to be able to put in a day?

JPG: I think so, yes.

SDC: And does it leave you feeling exhausted?

JPG: No.

SDC: It is an astonishing achievement. I suppose it's partly from a lifetime's habit of working.

JPG: Yes.

SDC: It seems you have a compulsion to work.

JPG: I like to feel that I'm contributing something, too. I don't like to shirk. If I cut you down to five minutes, I'd be shirking, d'you see!

One of Getty's secretaries, Carol Tier, came in and said: 'I'm terribly sorry, but I've got three businessmen in here, and Mr and Mrs Watson have arrived. I really don't know where to put them. I'm awfully sorry.'

I got up hurriedly and removed the tape recorder, saying: 'Not at all, Carol. It's not a problem at all.'

Getty sighed, 'No rest for the wicked.'

'I know,' I answered and went out towards the front door after being introduced to the Watsons.

'Well, it was an enjoyable hour,' Getty added as he said goodbye.

# 3
# ART AND THE MALIBU MUSEUM

Paul's butler, Bullimore, showed me in as usual. Bullimore, from the North Country, once worked for the Henry Fords in America and was in charge of thirty permanent staff at Sutton, including an assistant butler, footmen, upstairs and downstairs maids, cooks, assistant cooks, chars, and so forth. He insisted on lighting the fire while I was waiting.

At that moment Paul entered, Bullimore took his leave, and we began our most interesting conversation about his life as a collector.

SDC: I want to ask you about your art collection and the various ways in which you have found the objects you've collected. For instance, over there you've got that wonderful mirror with the boar on top and the two dogs converging on it – the gilt-framed mirror presumably from the eighteenth century. It's a fabulous thing. Where did you find it?

JPG: Actually, it came from the Duke of Westminster via Frank Partridge [a leading antiques dealer on Bond Street, London].

SDC: An expensive way round, but nevertheless one of the finest of its kind I have ever seen. The table underneath – the console table supported by figures and swags of fruit – where did you get that?

JPG: From Partridge, too.

SDC: He must have done some pretty good business with you. Yes, and these lovely pictures. You've had these pictures in this room since I've known you. Are they particular favourites of yours?

JPG: Yes, they are in a way. Of course, the price has gone up since I bought them.

SDC: You've got a Vernet here, I see.

JPG: I got that in Paris after the war, in the early fifties.

SDC: It's a pretty big picture showing a ship under sail in the middle and a sort of triumphal arch on the right. Then, there's the Avercamp snow scene of people skating on a frozen river. And the Ruisdael . . .

JPG: The Ruisdael was bought in 1938. I paid $6000 for it then. Now it would be worth $300,000–500,000!

SDC: Do you think that the rise in the value of works of art is a process that is going to go on indefinitely?

JPG: It's surprised everybody. A friend of mine was here for a few days, and he saw what the dealers had for sale in London. He said the prices are unbelievable. One picture I bought about four years ago and paid $70,000 for. They've got a similar picture by the same artist that is not in the best condition but the same size . . .

SDC: Who is the artist?

JPG: Lanfranco, and they're asking $200,000. Mine is in perfect condition.

SDC: You really always bought works of art for their own interest, fundamentally. You've never thought of them primarily as investments, have you?

JPG: I bought them because I liked them, but obviously, I like to buy them worth the money. I mean, I didn't buy anything just because some museum might have liked it. I didn't buy gory battle scenes or crucifixions.

SDC: You've been a very successful collector, obviously.

JPG: When I started buying art objects in the middle thirties a lot of my friends and business associates thought it was a weak spot in me – that I was really frittering away my money.

SDC: Whereas, in fact, you have been very, very shrewd in your purchases of works of art. Have you any idea – I suppose not – what your collection is valued at now?

JPG: I don't think you could duplicate it for over $100 million.

SDC: I would certainly agree with that. You hope to live

at your Ranch House in Malibu, California, when all the pictures have been moved to your new museum.

JPG: Yes.

SDC: So you'll be close to them, and you'll be able to enjoy them a little more than you've done lately. Do you go on collecting? You're an inveterate collector, aren't you?

JPG: Well, I say I'm through. I said I was through ten years ago. (*Getty's face cracked in an endearing smile.*)

SDC: But you keep doing it.

JPG: I said twenty years ago I was through; I said five years ago I was through; I said three years ago I was through; but, you're tempted. I've said, as most collectors have, that my collection is complete. Actually, the only thing I can say truthfully is that my resistance is higher now than it was twenty years ago. It's just like the paintings in this room. Now there's room for one more painting, and I suppose one could go there (*pointing to a gap between the windows*). But I don't really need another picture. So, my resistance is pretty high to a purchase of a work of art for here.

SDC: One over the celadon plate, that's the only place (*the plate, propped up, was pale green and Chinese of the Ming period*).

JPG: I don't feel that another painting is necessary unless it was something that bowled me over. If it was marvellous, wonderful, superb, and I must have it because it's been my dream to have something like that, then I might buy it. If it was just a good picture, though – you know, worth the money – I wouldn't succumb. Also, when you have a lot of empty space in a museum or a house, you're looking for art objects; you're actively in the market.

SDC: Do you ever weed out your collection? Do you ever feel as though you've got some marvellous things recently and you want to get rid of some of the stuff you got previously? Or do you always pass them on to the museum?

JPG: I've never weeded out anything, and I think I've only sold one picture, a Renoir. I bought it in the thirties. In 1946 a friend of mine made me an offer that was about a dozen times what I paid for it. For some reason or another, I sold it. It was the only picture I ever sold. Today, of course, it's worth a dozen times what I sold it for.

SDC: That was the only time you've been tempted to take a profit on a work of art?

JPG: Yes.

SDC: Have you been a collector all your life?

JPG: The first thing I collected was in 1912. I went to Japan and China, and I bought a little lacquerware. It was not of any importance, but it was the start of my collecting. The first work of art I bought was about 1930 at a Goldschmidt-Rothschild sale in Berlin. I bought a picture by Van Goyen for $1050. Then, the next sale I attended, the next purchase, was at the Thomas Fortune Ryan sale in 1932. I purchased some pictures by the Spanish artist Sorolla there. I really started buying after I rented Sutton Place in New York – Penthouse Number One. It belonged to Mrs Guest, who apparently was a relative of Churchill's. On their inventory were items like: 'One set of four chairs – Beauvais tapestry – $25,000. One table – $30,000.' I had to cover it with insurance in case of theft. I was taking responsibility for it. It was a very gorgeous place, and there were some very fine tapestries there. I took a year's lease, then another year's lease, then another, and then a five-year lease. Finally, starting in 1936, I began buying art objects of museum quality.

SDC: In 1936 you'd have been forty-odd years old. You have very catholic taste in art. In this room they're all old masters, such as Van Goyen, Ruisdael and Avercamp – and a Cuyp, I think. Yet, when you were showing me around the house years ago, I noticed you had pictures by Renoir, Pissarro, and such in your room. You've never felt any inconsistency in having these modern artists?

JPG: Quality is quality.

SDC: Which are the favourites of the modern paintings you've got?

JPG: Actually, I think my favourite is a Pissarro landscape I've got upstairs.

SDC: Who do you think are the great artists of the last century?

JPG: I like the Impressionists – Renoir, Degas, Pissarro.

SDC: Of course, you have moved a lot of your paintings to Malibu.

JPG: Nineteen pictures. I sent them about three years ago.

SDC: So the bulk of your collection is there, but the picture of the Marten Looten – the Rembrandt – you gave that one to the Los Angeles Museum, didn't you?

JPG: Yes, twenty years ago. Then I decided that, instead of giving to other people's museums, I'd establish my own museum – give to my own.

SDC: And you also have important pictures away on loan. For example, *The Madonna of Loreto* in the National Gallery. I presume anytime you wanted it you could always ask to have it back.

JPG: Yes, it'll have to go to the Malibu Museum. It'll have to leave the National Gallery. It's been there nearly three years now.

SDC: And there's also the self-portrait of Paolo Veronese.

JPG: Yes, that'll have to go to Malibu too.

SDC: Have you enough things all over the world to fill your very big museum?

JPG: Yes, I think I have.

SDC: So you're not looking for more purely to put in the museum?

JPG: My attitude is more or less the attitude of a man who listens to a comedian and says: 'Louder and funnier'. In other words, I'm saying: 'Now try hard and make me laugh. I'm not laughing spontaneously.' If it forces a laugh out of me, I can't resist it.

SDC: Are there any works of art you covet, any particular artists whose work you rather hanker after? I don't think you have a Vermeer...

JPG: No, I haven't. I'm reconciled that I can't have the works of every artist. I always like to have a good Greek original statue.

SDC: Did you see the supplement in the *Observer* [23 January 1974] on your Malibu Museum? It seems to have caused a great excitement when it opened to the public.

JPG: We had capacity crowds there; they were lined up for 5 miles [8 kilometres]. The police were out in force.

SDC: Well, it's not surprising. It's a very exciting new venture. A lot of people seem to be very startled by the warm colourings of the Roman period. This was your own idea, was it, to keep the colours of the period?

JPG: Yes, to make it authentic.

SDC: Some people say, of course, that the colours are so dramatic, the place so dramatic, that it almost took the limelight off the collection inside.

JPG: Pictures weren't created and statues weren't created to be displayed in museums. They were created for decoration for homes, palaces, churches, temples. They were part of the decoration of a building. Actually, a picture looks better hanging in this room than it would in a museum. For example, the Vernet. These pictures, I think, look better in a home than they would in the National Gallery. Do you agree with me?

SDC: Entirely, yes. Of course, it's not everybody who is fortunate enough to have a setting like this to show off their pictures, and most people have to rely on what they can see in museums.

The J. Paul Getty Museum, however, is your main legacy to the art world. How did you come up with the idea of recreating a villa from Herculaneum? When did you start thinking about all this?

JPG: Well, I wrote a book in collaboration with Ethel leVane in 1955 called 'The Journey From Corinth' [actually part of *Collector's Choice* – see page 120].

SDC: I read it.

JPG: The villa in Herculaneum, the Roman Villa, is the one in my story.

SDC: Did you see the foundations of this one at Herculaneum? How much of it survived?

JPG: I was always interested in Pompeii and Herculaneum. Of course, when I wrote the story I had no thought of building a big museum or even having a museum, much less recreating the villa. The thing about a museum is that they're very expensive to build and very expensive to maintain.

SDC: What did this one cost you?

JPG: The building cost $17 million so far, and it will cost about $1 million a year in maintenance. We have a pretty big payroll and other expenses, so you'd be getting off cheap for $1 million a year.

SDC: And it will go on getting more expensive every year with inflation. How have you catered for this increasing expenditure? Are you setting up some sort of trust fund that will grow with the years?

JPG: You have to have an endowment fund for the museum.

SDC: You feel that it will, for the foreseeable future, cope with escalating costs?

JPG: I should hope so, but of course if it doesn't, you've got to charge an entrance fee.

SDC: When did it occur to you to build a museum to house your collection?

JPG: Well, as we've said, I planned to go back and resume residence in Malibu, and my collection got too big for the wing of the house. I didn't want to use part of the house as a museum, so I decided to build a separate museum near the house. I made that decision over three years ago.

SDC: You made remarkable progress in just three years. How did you start translating the Herculaneum villa into bricks and mortar?

JPG: As you may know, Herculaneum is different from Pompeii. Pompeii was buried only at a depth of 15 to 20 feet [4.5 to 6 metres] in rather soft ashes, so it was easy to excavate, comparatively speaking; whereas Herculaneum was buried under mud that had hardened into tufa and was 50 to 100 feet [15 to 30 metres] deep. In Herculaneum nothing was taken away by the Romans after the disaster because it was too deeply buried. However, they did take away a lot of goods from Pompeii because it was practical to do so. The Herculaneum villa was explored in the eighteenth century, and they built galleries. A good Swiss engineer in charge of the work, Weber, left a very good set of engineering drawings of the peristyle garden in the house attained by these galleries, fortunately for us. Works of art found in the villa have become very important contents of the Naples Museum. Mauri, who was director of the excavations at Pompeii and Herculaneum for about twenty or thirty years, said it was the largest and finest building ever found.

SDC: Did they have the exact measurements of the whole site? Were they able to find the whole site?

JPG: Yes, they had the measurements. I made as good a copy as I could, although probably there are some slight changes.

SDC: Did it have a name in Roman times?

JPG: We don't know for certain who the proprietor of the building was. It was an enormous villa.

SDC: The 'J. Paul Getty' of 100 BC or AD 62?

JPG: We think it was probably the father-in-law of Julius Caesar, Calpurnius Piso. It was probably him. Cicero speaks of him in one of his speeches. He was greatly interested in epicurean philosophy, and they found in the library a great number of books on epicurean philosophy. There are certain other reasons to believe he was the owner, but we can't be certain.

SDC: Very interesting. So you knew the exact outlines of the original villa and the probable owner. Then, did you discuss this with the various architects, or had you already had somebody in mind?

JPG: We tried to minimize the modern architect's contribution to it by making as fair a copy as possible of the Roman architecture. We got a man named Professor Neuerburg who was connected with the work done by the director of excavations at Pompeii and Herculaneum for a while. He's the archaeologist we retained as a consultant, and I think he has been very instrumental in keeping it authentic.

SDC: But he was not the architect of the new building, presumably?

JPG: We had an architect who complied with the codes, but they were making a copy; they did not originate the design. So, I say the architect is unknown. We don't know who the architect was.

SDC: The architect being a Roman?

JPG: A Roman, yes.

SDC: But, of course, there are tremendous constructional problems that only a modern architect can use to advantage.

JPG: Yes, you've got all the requirements of the building code too regarding staircases and strength construction. It's in an earthquake belt as most of California is. Because of that, the building code calls for certain strengthening of structures.

SDC: Does that situation worry you?

JPG: I got a letter from Jean Dixon, which I could show you. She is a famous fortune-teller/psychic. She says that most of the works of art at Malibu will be destroyed.

SDC (*nearly jumping out of the chair*): Oh? That's a very uncomfortable prophecy. Is she to be relied upon as a clairvoyant?

JPG: Well, she's been very successful.

SDC: I'd like to see it.

*(Getty goes in search of the letter and returns, handing it over.)*

SDC: This is Jean Dixon writing, is it, on August the twenty-seventh of this year, 1973? (*I read the letter aloud.*)

Dear Mr Getty,
You are planning a great exhibition at Malibu in California to house your art treasures. Please consider. It has been revealed to me that many of your finest paintings will be lost or destroyed by a natural disaster in or near Malibu. For the world to lose such treasures would be a great tragedy, especially those of the Greco-Roman era and the French eighteenth century. To avoid this, I urge you to move them inland, possibly to Houston, Texas. Please know, Mr Getty, I use my gift or prophecy to try and help others. Your fore-ordained mission, Mr Getty, is to share your great wealth for the good of humanity, and only its good. There is great goodness in the beauty of art, as there is great goodness in the beauty of children – a subject I believe is of great interest to both of us.
Respectfully,
Jean Dixon

SDC: I see you replied on the twenty-sixth of September 1973.

Dear Mrs Dixon,

Thank you for your kind letter of the 27th August, 1973. You may be right, but I hope you are not. In any event, I will bear in mind what you say. I have much pleasure in enclosing a token donation towards the worthwhile work you're doing.

With all best wishes,

Yours sincerely,
Paul Getty

SDC: The organization she writes from is called 'Children to Children, Incorporated'. Well, do you pay any attention to a prophecy of that kind?

JPG (*wearily*): How can I? I'm finishing the museum.

SDC: Do you think she's predicting that there will be an earthquake there?

JPG: An earthquake or a tidal wave.

SDC: Are you superstitious at all? Most of us are!

JPG (*smiling*): Not particularly, but then, as you say, most of us are. I don't go out of my way to walk under a ladder.

SDC: No. If I got a letter like that, I think I would dismiss it as the letter of a crackpot.

JPG: Well, I would, except that she's so famous.

SDC: Is she? Has she made any other prognostications that have come off?

JPG: Yes, she predicted some years ago that Nixon would be elected President, that he would be elected to a second term, that he'd have a dire scandal, but that he would come out of it all right.

SDC: This could be regarded as a rather shrewd political appraisal.

JPG: Yes, but she's predicted many things.

SDC: So what could you possibly do at this stage, even if you really believed it? You've built your museum, as you say, and . . .

JPG: An earthquake's as much as you can get.

SDC: Let's hope and pray that she's wrong about this prediction. This sort of thing doesn't make your breakfast feel any better, does it, when you open the mail?

JPG: No.

SDC: Well, you've disconcerted me. I don't think I'd like to see a letter like that prophesying that all your collection is going up in smoke. It would be a tremendous loss to the world. You have collected some of the greatest masterpieces. You've got Rembrandt of St Bartholomew – the one I always describe as 'eating peas off his knife'. And you've got Rubens of Diana there, haven't you?

JPG: Yes – departing for the hunt.

SDC: Is there any method whereby you can have a rapid evacuation of the principal works of art? If one had some kind of warning, the works of art could be dismantled rapidly and removed within hours to another site; or would there never be enough warning for that?

JPG: They're working on strain gauges of these faults in California to see if we couldn't anticipate earthquakes a few days ahead. Of course, if we got enough notice, we could remove the objects; but as far as I know, our particular location is not in a dangerous zone. The house wasn't damaged in the great earthquake of 1933, which was more or less at Long Beach. The worst part of the damage was around Long Beach. Nor was it damaged in the great earthquake of twenty-three years ago. The modern house has been there for fifty years, right though two major earthquakes, with no damage. So I don't believe it's in a particular earthquake area; it's not on a fault as far as we know. As a matter of fact, there's been very little damage done in Santa Monica or Malibu by earthquakes in historical times.

SDC: I saw the museum during its construction. I'm astonished by the fact that all has been done within three years, since you first decided to do it.

JPG: We had a very good builder; the workers have been pushed. I expect it to be completed in January. I'd like to be there for the official opening. The unofficial opening was without any ceremony.

SDC: What sort of official opening have you in mind for it?

JPG: They have a public figure to open it. (*Getty did not volunteer the identity.*)

SDC: And you'll be there, that's marvellous. I hope you'll let me come along and see you do that.

JPG: Yes.

SDC: It would be a great moment. Will it be open to the public free of charge?

JPG: Yes.

SDC: I think it is much better if people can go freely to a museum. You've seen the Huntington Library in California. I don't know how many people visit it in the course of a year, presumably a great many.

JPG: Is it free?

SDC: It was when I went there. I remember, apart from Gainsborough's *Blue Boy*, an amusing letter written by George III to Lord North, his Prime Minister. It said that if the American colonies persisted in their present course of action, he wouldn't be responsible for the consequences. Well, the consequences have been quite considerable.

JPG (*laughing*): Yes, they have been.

SDC: What about the content of your collection?

JPG: I suppose the collection reflects objects I was interested in. I haven't tried to make a set collection or formal collection. I've left that for the big national institutions. I've not tried to cover every school and every period and every branch of art.

SDC: It's rather comparable to the Frick collection in New York, isn't it, in that respect, of a private collection of a single individual?

JPG: For example, I have no Oriental art, with the exception of carpets. Not that I'm not interested in Chinese or Hindu art. I just didn't have time, money, or space. I can't have perfection like the combined British Museum, National Gallery and Victoria and Albert Museum.

SDC: No, this is a personal reflection of your taste in art, the things that interested you most and that you have accumulated during a very acquisitive life.

JPG: Yes, and the emphasis has been on decorative art, so-called, because I feel that a piece of furniture can be a work of art just as a painting can be, or a sculpture even more so.

I think a sculpture can be more a work of art even than a painting.

SDC: Because it's three-dimensional?

JPG: I suppose in a way – it would be more lifelike. It can be truer than nature.

SDC: You remember Michelangelo when he finished the statue of Moses. He struck it with a hammer and said: 'Now, speak, damn you,' or words to that effect. Meaning that it had got to the stage of almost being alive. You feel that sculpture is really more living than painting?

JPG: Obviously you don't have a landscape in sculpture, but I think that if you want a portrait, a statue can give you a better portrait than a painting could.

SDC: Would you say that in relation to the great self-portraits by Rembrandt? Don't you feel when you look at the sad, melancholy faces, reflected no doubt from the mirror of Rembrandt, that they are as close to the man as you could possibly get? That if they were in cold marble with sightless eyes, they would not in fact convey so much of Rembrandt's personality?

JPG: Let me remind you that in ancient times marble was painted.

SDC: I know. One wonders, in fact, what the statues must have looked like when painted.

JPG: Apparently, the Greeks thought that marble was the supreme work of art.

SDC: It's very intriguing you say that. The curious thing about statuary in marble is that it is not really thought of as a tactile form of art in the sense that jade is. The Chinese, when they have in front of them a piece of jade, think of it not only as a thing to look at but a thing to feel. There's a great sensitivity in jade, and they particularly like nephrite jade, which has a wet feel to it. Rich mandarins kept jade buffaloes on their tables as a kind of 'Alka-Seltzer', a sort of digestive tablet; they would stroke the smooth surfaces. It seems to apply only to carving in jade. One wonders why it doesn't apply to a carving in stone or marble.

JPG: I think it's just my personal taste. I think if I had a choice of taking the best paintings in the museum or taking the best statue from the Athens Museum, I'd rather have the best statue in the Athens Museum than I would the best painting in the Vatican collection. It would be a hard choice, let's say, between the *Mona Lisa* and the Soldier of Marathon in the Athens Museum, but I think I'd rather have the Kouros.

SDC: Well, this is a bold point of view, I think. Very interesting and, of course, you have got a marvellous collection of marbles yourself. Are there any particular works of art, though, that you would still like to get to fill up some part of the museum or for your own private use?

JPG: Oh, I don't know. I would always like to have a good original Greek statue.

SDC: They are getting harder and harder to find, aren't they?

JPG: Yes.

SDC: You've got that book, I think, on *Great Private Collections* [by Gerald Cooper; Weidenfeld & Nicolson, 1964], which has your collection in it. Remember, I got you to lunch at the club in London with George Churchill, who has his collection in that book also. He had the bronze deer he always said was done by Myron, a fifth century BC bronze. I managed to get it when he died. You know everything he had was sold.

JPG: It's very fine.

SDC: Do you know he got that for £150 in Bologna when he was looking around for works of art, and he'd been having a very bad day. Suddenly, he saw this bronze deer on top of a cupboard in an antique dealer's shop. He asked what it was, and the owner answered that he was going to show it to a professor to find out. George said: 'Well, I'm a professor, you can show it to me.' This was the bronze deer – the fifth-century Greek bronze deer – that became the centrepiece of his collection. He had that Cretan bull with the acrobat on the back also, which he bought in two parts from Spink for £100. When he died it was valued by Peter Chance of Christie's for £75,000. The British Museum had to pay £50,000 for it. So you are in the same position; a lot of things you acquired in the early days were bought very cheap.

JPG: Yes, it's rather amazing. About twenty years ago a dealer – I won't say where – showed me a life-size marble statue of the crouching Venus with the head on it, and I bought it right away. At that time it was about $12,000. The head was there, but it was positioned improperly, and I had it positioned properly. The dealer had bought it from a man who had bought it from another dealer who dealt in garden objects; the statue had been out in a garden. I don't know what the garden dealer got for the statue, but I'll guess that he might have got $500. Then, the city dealer might have paid $5000 for it, and then sold it to me for $12,000. Actually, though, at the time it was easily worth $250,000 to anybody who knew anything.

SDC: Even then?

JPG: Even then, because I have a catalogue of the Prado Museum, and they have a crouching Venus – they have it as the frontispiece of their catalogue. They have quite a few $10-million paintings, but they had that as the frontispiece. The Louvre has a crouching Venus, but neither the Louvre nor the Prado's crouching Venus has a head. They were headless, and they don't have a Cupid either. Mine has the head of Cupid.

SDC: Is this fifth century BC?

JPG: It's third century. Mine is a Roman copy, but it's life-size. But, it's a rarity that it has the head and the Cupid, and it's in good condition and intact! If you have the Greek original, which is one of the most famous statues of antiquity, it would probably be worth today at least $10 million! A good Roman copy of it today would be worth possibly $500,000 to $600,000.

SDC: So the one you've got would be worth about $500,000 or $600,000 now?

JPG: I would like to think so, yes.

SDC: But the Greek original is lost, presumably?

JPG: The Greek original hasn't come down to us, but the

point is that none of the previous owners knew that they had the crouching Venus!

SDC: So your shrewd eye got hold of a bargain.

JPG: Yes.

SDC: Do you feel that people really can go on – private collectors can't go on buying at such high figures, can they?

JPG: I wouldn't think so.

SDC: Of the great works of antiquity that you have seen in reproduction, which are your favourite ones? Would you say that the Charioteer of Delphi was one of the finest?

JPG: Of the ones we have, I would say that the Charioteer is very good. The Zeus is very good. They're both bronzes. But probably the most valuable work in the Athens Museum is the one they found not so many years ago – twenty-five or thirty years ago – the life-size soldier. The date is before Marathon, about 500 BC.

SDC: Have you seen it?

JPG: I saw it at the Athens Museum. I think that was the finest thing they have in the museum.

SDC: Did you see the bronze Charioteer?

JPG: Yes, I saw that at Delphi. It's very fine.

SDC: Obviously, this is one of your main interests in art, isn't it, classical sculpture?

JPG: Yes. I saw the famous Hermes of Praxiteles at Olympia, as well.

SDC: Is it as fabulous as it looks in the pictures?

JPG: Yes.

SDC: How long ago was it that you saw these things in Greece?

JPG: Let me see, when was it I was in Greece – 1953 or 1952 – twenty-one years ago. I had been there before, in 1913. I've always thought that the great period of Greek art has never been matched since.

SDC: Of course, it only applies to sculpture. There's very little painting.

JPG: None of their paintings have come down to us except in frescos, and of course, we don't know whether that was a fair sample of their easel paintings or not.

SDC: You don't feel then that Michelangelo was a greater sculptor than Praxiteles or Phidias?

JPG: I don't, no.

SDC: To a certain extent he was a follower of those anyway, so he wouldn't have the same originality as the first people who did this.

JPG: I think it was Michelangelo who complained that they never compared his works with his contemporaries – the ancient Greeks.

SDC: Yes, quite right.

JPG: I thought that was rather unfair. Also, in one case, Michelangelo was asked to restore an ancient statue – to put on the arms – and he said he wasn't capable of doing it.

SDC: Marvellous humility. You've also shown a great interest in furniture. You have some famous pieces, of course, at Malibu. Even in this room – that cabinet with the little drawers is obviously a favourite of yours. (*I indicated a lacquer cabinet near the door to the hall with a large number of small drawers in red and gold.*)

JPG: Yes, it is. I got that at a dealer's exhibit at Grosvenor House about thirteen years ago.

SDC: So you've gradually been topping up your collection all the time?

JPG: Yes. I'm interested in French eighteenth-century furniture, Italian Renaissance furniture, tapestries, and Persian carpets.

SDC: How did you get the Ardabil carpet? This is one thing I've always rather envied you because there is one in the Victoria and Albert Museum. Presumably, there were two made for the mosque. This was made about when, 1515 or so?

JPG: Yes. Well, I had seen it before in an exhibition. I saw it about 1936 in London in an exhibition of Persian art. I got word that it was owned by Duveen. I tried to buy it from him, but he didn't want to sell. I finally bought it from him right after the war broke out.

SDC: What did you have to pay for it?

JPG: I paid $68,000 for it.

SDC: That was quite a lot at the time, of course, but certainly not when I think about it now. You must have been a bit fed up about the Titian, having put down all that money for it. [Getty had bought the Titian of Diana and Actaeon at Christie's for $4 million, but the Committee on Exports of Works of Art had withheld a licence for six months and then opposed the export.]

JPG: Well, of course, the museum has lost the interest on the money.

SDC: I know. That seems rather unfair to anybody, really.

JPG: It did seem a little unfair to me because of the length of time. You see, it had been assumed that they would take ninety days. In previous instances I don't think they've ever taken longer than ninety days, and I had made arrangements that I was going to pay for it in ninety days. Instead of ninety days, though, they got a year!

SDC: That's a lot of money – a lot of interest on $4 million.

JPG: So the museum lost the interest on the money for nine months; it was about £100,000.

SDC: Rather a sickening experience. You must have been keen on that particular Titian of Diana and Actaeon, yet you didn't seem to have taken anything like the same interest in Rembrandt's Titus when that came up. Now, this is the sort of picture I should have thought would have appealed to you even more. Did you go for it at all?

JPG: No, I didn't.

SDC: Art is a very subjective thing, isn't it? You have to decide entirely what it does for you. You were saying that the things that have given you the most pleasure are the marbles, the Greek and Hellenistic marbles you have bought. Which are the ones you attribute most importance to?

JPG: I think my statue of Hercules.

SDC: You feel you've rounded off your collection for the museum, and you will undoubtedly go down in history as a public benefactor of art. Have you also done a great deal in the sphere of charity, setting up foundations of any other kind? Some people, like Sir Isaac Wolfson for instance, have given a great deal of money to medical research. Nuffield did, too. Has this ever been a principal interest of yours?

JPG: No, it hasn't. I thought that many other people have done that. It's certainly a very worthy way to spend money, but art has a place too.

SDC: Very much so. In fact, it could be said that art has a more important and enduring place, I think.

JPG: I can see myself, when you're thinking about philanthropists, that most people donate art to an existing museum so they're not burdened with the maintenance of the museum. But when you build something like I'm building, and you think about the maintenance...

SDC: Terrifying.

JPG: And it goes on; the life of the museum might be hundreds of years.

SDC: This museum of yours is going to last, we hope, longer than the one built at Herculaneum, and if it lasts for a thousand years...

JPG: I wonder if that fortune-teller was acquainted with the history of the prototype.

SDC: Maybe the same fortune-teller who called out to the owner's son-in-law: 'Beware the Ides of March', because she was certainly accurate!

JPG: What time will you be here tomorrow?

SDC: Three o'clock, if that suits you, of course.

JPG: That'll be alright. Thank you and goodbye, Somerset.

SDC: Goodbye, old chap.

# 4
# THE EARLY DAYS IN OIL

Altogether, it was a good afternoon with Sutton looking at its best in bright sunshine. After greeting me at the front door, Paul told me how pleased he was at how the book was progressing. We walked to the drawing-room, assumed our chairs, and began talking.

SDC: Let's talk about your family, if we may.
JPG: Of course.
SDC: Your mother was Scottish, wasn't she?
JPG: Yes. Her mother was a McPherson, and her father was a Risher. He was Dutch, so there was some Dutch blood. My father was pure Irish.
SDC: I know the story about your father's family. They were the ones who went from Ireland first to France and then to America at the end of the eighteenth century, is that not so?

> **On Gettysburg**
> A cousin of my grandfather, by the name of James Getty, came to America in 1780. He was a Presbyterian of Irish descent with a Scottish mother. From the heirs of William Penn, he bought the land where Gettysburg, Pennsylvania, now stands. In 1786, James Getty had a plan of the town made, hence Gettysburg.

JPG: Yes. John Getty was born in Gettystown, Londonderry County, Ireland, about 1760, and he died about 1830. He was the great-grandson of John Getty of Cullormore, Ireland (later known at Gettystown), who was mentioned in the tax rolls as living there in 1663. A cousin of my grandfather, by the name of James Getty, came to America in 1780. He was a Presbyterian of Irish descent with a Scottish mother.

## The Early Days in Oil

From the heirs of William Penn, he bought the land where Gettysburg, Pennsylvania, now stands. In 1786, James Getty had a plan of the town made, hence Gettysburg. In 1800, Adams County was formed and Gettysburg became the county seat.

SDC: Gettysburg has, of course, acquired great fame in America. Did any other member of the family, other than your father, have a particular aptitude for making money? It seems that your father was the first.

JPG: I don't think so. Here's some background information about my father (*reading aloud*): 'George Franklin Getty was born near Grantsville, Allegheny County, near the mountain slopes of western Maryland, on October seventeenth, 1855. His parents, John Getty and Martha Ann Wiley Getty were, as well as their ancestors for several generations, all settlers in the region and tillers of the soil' – except for my Grandfather Wiley, who was an influence in the community both as its pastor and teacher. By teaching school in the winter term, my father secured funds to pursue his studies at the university for the rest of each year. He graduated from the Scientific department of Ohio Northern with honours on July tenth, 1879.

SDC: The fact that your father was successful – and very successful by the standards of the time in the oil business – was pretty decisive in your going into it, wasn't it?

JPG: Yes. My father was a remarkable man because he was only six years old when his father died, and his mother was an invalid. He had two younger sisters and a younger brother, so he was head of the family at six.

(*Getty then drew my attention to an illustration in the book he was holding in his hand. I got up and looked over his shoulder.*)

JPG: This is my father's uncle.

SDC: Senator William Reid Getty. This was taken about 1895. Yesterday, you gave me a rather interesting photograph of the uncle who was a general on the Union side.

JPG: Yes, that was the same generation.

SDC: He was a very striking, very fine-looking man. Was he a successful general? I mean, not all generals are.

JPG: I think he was; he was supposed to have won the Battle of the Wilderness. The Gettys were a very handsome people – men and women both. Except my father; he was nice-looking, but he wasn't handsome. He had been in partnership with his cousin, John Getty, who was very handsome. My mother told me of the time she was in the office and a woman came in and wanted to see Mr Getty. My mother was alone in the office and asked: 'Which Mr Getty do you want to see?' The woman replied: 'The handsome Mr Getty.'

SDC: Not very complimentary.

JPG: They sent my father out.

SDC: What happened to the handsome one?

JPG: He died years ago. I was in a hotel in Tulsa about 1914. I was in the lobby and was being paged. A messenger boy came up and said: 'The United States Senator Pine would like to see you.' So, I followed him to where the Senator was, and the Senator excused himself. He said: 'I thought you were John Getty. I heard them paging Mr Getty, and I just took a chance that you were John Getty, but frankly speaking, you don't look much like him.' I was about twenty-two. I can remember this very well. He said: 'John Getty was the handsomest man I ever saw.'

SDC: Your mother sounds a terrific character, obviously a strong character.

JPG: Yes, she was.

SDC: Would you share some recollections?

JPG: Well, my father met her at Ohio Northern University. They were both students. He was the editor of the school paper, and she was his assistant editor. They got married just after they graduated; but he owed $100, and she didn't want to start married life in debt, so she paid off his debts. A $100 was a lot of money in those days.

SDC: What did they marry on?

JPG: Well, I think he had very little, and she didn't have much more.

SDC: What job did he take?

JPG: He was teaching school. He kept on teaching school and saving money and went to the University of Michigan, where he graduated from law school. Then he went straight into practising law and was elected a circuit court commissioner for two years. That was sort of a judgeship. Finally, he decided to go to Minneapolis. He said, by all accounts, that it was because of his wife's health; but that didn't seem very logical to me because Minneapolis has one of the toughest climates in the United States!

SDC: That can't have been very good for her health.

JPG: No, I wouldn't think so.

SDC: Any brothers or sisters?

JPG: I had one sister, but she died nearly three years before I was born, and so, of course, I never remembered her.

SDC: I understand that your mother was forty when she had you and that your father, thirty-seven at the time, regarded it as something of an Old Testament miracle that she should have conceived again.

JPG: Yes. My birthdate is December 15, 1892, and I understand that it was sixteen degrees below zero when I was born! I was baptized in the Methodist Church as Jean Paul Getty, the 'Jean' a reminder of the first John who had been in France during the Revolution.

SDC: When did you first travel to Oklahoma?

JPG: When I was twelve we visited Indian territory. I was disappointed to find there are no Red Indians. Actually, it wasn't known as Oklahoma yet. My father was acting as legal advisor to the North Western Life Insurance Company. They had a policyholder who was making a claim against the insurance company. For the time, we settled down in Bartlesville.

SDC: I've been reading a good deal from your own autobiography [*My Life and Fortunes*, 1963]. There seems to have been a remarkably entertaining character in Bartlesville, a highwayman called Henry Starr. Do you remember him personally?

JPG: Yes, he was quite a character . . . 'a man of his word'. He reformed and was paroled, and when my father started in the oil business, he gave Starr a job. Of course, he was the centre of attraction. He had murdered quite a number of people, but they couldn't prove it. There were never witnesses.

SDC: I like the splendid story about Starr's constantly sticking up Bartles, for whom the town was named and who never had much money on him. Starr resented this and finally said to him: 'If you don't have a decent amount, $5000, next time, I'll shoot you in the guts!' Is it true that Bartles actually took the trouble to go around with the money on him after that?

JPG: Yes, that's so. That's the story.

(*Getty had brought in with him his privately-printed* History of the Oil Business of George F. & J. Paul Getty. *He got out a pair of tortoise-shell-topped spectacles, which he balanced with some difficulty on his nose, and read rather rapidly from the heavy, green-bound volume on his knees.*)

In 1896, father had a serious attack of typhoid fever, and for a few days his life was despaired of. His recovery was slow, and for several years, he more or less retired from active business.

(*Getty stopped reading and looked up.*)

JPG: I can remember his attack of typhoid fever. That was before my fourth birthday. He was ill a few months earlier, in the late summer of 1896. He had a favourite doctor who was a friend of his, Dr Bell, a very fine man. I can remember Dr Bell calling every day, twice a day, to see how he was getting along. I can remember the good doctor taking me for a walk and telling me that it was possible that my father might not recover and that I would have to give support to my mother.

SDC: This was before you were four? You've got a good memory.

JPG: Oh yes, I remember that. Of course, it was a powerful experience.

SDC: He recovered from that?

JPG: He recovered, yes, but he wasn't able to work. He was fairly fit physically, but he was unable to work and the doctors didn't seem to be able to help him. That's when he took up Christian Science, which gave him courage and helped him a great deal.

SDC: But your mother never took it up?

JPG: No. In fact, they had arguments about it. She had a low opinion of it.

SDC: You are a Methodist?

JPG: Yes, I went to Wesley Methodist Church.

(*Getty started reading again.*)

Now, in 1900 – that was when I was seven years old – father was living with mother and me, his only child, in a flat at 1678 Hennepin Avenue. That was in Minneapolis. He was in good financial circumstances with $47,160 in a checking account in one bank and accounts in two other banks. He kept one or two horses. The records show $20 a month for horse board and $2 for shoeing. He had his suits made by Pease, then considered the best tailor in town, and according to an old bill, $65 was the price of a lounge or business suit.

JPG: They were the good old days, weren't they?

SDC: Yes, yes.

JPG: Men wore high stiff collars then, and the suits had short, straight coats that were buttoned very high. The trousers were long, without any crease, and the turn-up or cuff at the bottom was, as yet, unknown.

> **On religion**
> I am a Methodist.

SDC: When did he actually enter the oil business.

JPG: In September of 1914 in Tulsa, Oklahoma.

SDC: I know you'd been a roustabout on your father's rigs and so on. Had you studied geology or had any training?

JPG: Yes, I'd studied geology formerly. I hadn't studied at school, but I'd read textbooks and I believed in it. Many at that time did not believe in it.

SDC: You've referred to the old 'luck' in all this, and obviously, there was some luck; but equally, do you feel there is a particular kind of instinct? You've referred to Barnsdall's remark: 'You can sniff it even if it's 3000 feet [900 m] underground. That is the way to find oil, you just sniff it.' Did you ever feel this?

JPG: No, I never did. I trusted more in geology. I remember a man who was a manager for a major oil company, and he didn't believe in geology. This was when I first started. I was talking to him, and he told me about the lease that the company had: 'It was right square in the middle of the sinticline.' It was a mispronunciation. He thought it was good; he said it looked good. I didn't because the syncline is where you expect to get water, and the anticline is where you get oil!

SDC: Anyway, you knew a good deal about it.

JPG: It's obvious when you're drilling off-shore – drilling in the North Sea – you're completely dependent upon geology. How can you sniff oil? How could you tell from the lay of the land when there is no land? Would you say that oil was there because the waves were a particular shape?

SDC: You were held up by gunmen or bandits in the early days. You said you had a letter from a local council asking if they could put up a notice saying: 'This is where Paul Getty was held up.' Could you describe the episode?

JPG: Well, it was going from the town of Cushing, on the train to Drumright, which was the principal town. It was a very wild country at that time, about 1914. People were frequently relieved of their purses, so most people didn't carry much money with them. They only got a few dollars out of us, that's all. There really wasn't much to it. It was more or less considered one of the hazards. When I first went to Drumright, I met a friend there. I noticed that when we were going round a corner, he would stop and take off his hat and put it out before he turned the corner, getting close to the building. I said: 'Why are you doing that?' He answered: 'Well, I don't want to be hit over the head or shot. I put out my hat to see if there is anybody there.'

SDC: This period of your life was a very colourful one.

JPG: Yes.

SDC: Did you enjoy the business of oil drilling?

JPG: Yes, very much. I like working with my hands.

SDC: How long did you do that?

JPG: I started in 1904 and finished about 1919.

SDC: And what was the work like?

JPG: It's pretty scary work. A friend of mine – I got him a job on a rig – worked one day and gave it up. When I asked him why, he said it was because he didn't like looking up and seeing ten tons over his head all the time. The tools, you know. The drilling bits, and so forth, and the drill pipe, and the blocks – big blocks that go up the derrick.

SDC: And what are you doing underneath, exactly?

JPG: Twisting – the drill part revolves. There's a bit at the end of it.

SDC: And you're guiding this into the ground, is that it?

JPG: Guiding it into the ground, and then the big blocks are moving up and down over your head all the time. Sometimes you have to climb to the crown on the derricks, about 122 feet [37 metres].

SDC: That is a very great height. It's rather like getting to the top of a mast in the old sailing ships. Did you have to go up there a great deal?

JPG: Yes. I think the insurance rate is about as high as it is for lion-tamers. Actually, it's just as dangerous, or more dangerous, than coal mining.

SDC: I don't think the public is aware of that. Of course, they realize that drilling off-shore in the storms and weather conditions of the North Sea must be very dangerous and difficult. Is it a very dirty occupation? I mean, do you get covered with oil?

JPG: Yes, you do.

SDC: How many hours a day did you do this?

JPG: Twelve hours a day, seven days a week. And the rest we had was between wells.

SDC: And you kept this up for several years, working as a roustabout. Was this purely to get experience or as part of the firm?

JPG: I worked as a roughneck and a roustabout, both.

SDC: What is a roughneck?

JPG: A roughneck is a member of the drilling crew. A roustabout is one who handles pipe and does whatever is to be done apart from actually drilling the well.

SDC: I suppose there are certain skills in this business that are more highly paid than others. What is the most highly paid?

JPG: A driller. A driller can cost you, maybe, the price of a well. If he makes a mistake, you can lose the well and the million dollars that well cost. So, you're dependent on his accuracy, good judgement, and skill.

SDC: I was fascinated by the story that when one of the bits jammed – you used to lose bits sometimes and you would have to fish for them – you had a tombstone made up. What happened exactly?

JPG: We get what we call 'sidetracking', and we wouldn't be able to sidetrack successfully. So, the idea occurred to me to get some granite and make a whipstock, drop it down the hole, go in with the bit, and be deflected past the fish. The part you're trying to sidetrack is called the 'fish'. It's part of the drilling string, which is unfortunately left in the hole. It worked out rather successfully. They used to call them 'Paul Getty specials'. The granite whipstock was made by the tombstone-maker.

SDC: They were happy days of yours, in the oil fields, weren't they?

JPG: Yes. Unfortunately, a long time ago.

Paul's mother, as stated earlier, was a very strong character. In researching some of Paul's private papers, I came across this letter from her to Paul handing over her shares at Christmas 1933:

> This offer shall remain open to and until 12 o'clock noon, December 30, 1933, and if not accepted by you in writing on or before that date and hour shall be considered as withdrawn by the undersigned and shall be wholly terminated and at an end.
>
> Very truly yours,
> Sarah C. Getty

This is a tough mother!

# 5
# ROMANCE AND MARRIAGE

Paul always had an eye for the beautiful in art and the beautiful in women. All five of his marriages were to lovely ladies who were considerably younger than he was. I often thought that his best relationships with the opposite sex were with special friends, such as Penelope Kitson, Margaret, Duchess of Argyll, and other charming and interesting companions. We talked at great length about his philosophies on women and marriage.

SDC: Was your parents' a very happy marriage?

JPG: Yes, it was. They celebrated their golden wedding anniversary. She went on to live to be eighty-nine minus one month.

SDC: In the marriage stakes, I'm afraid, rather like me, you've been a sprinter rather than a stayer.

JPG: Yes.

> **On marriage**
> I think it is very often the case that a man that fails in everything else, does not fail in his marriage.

SDC: Do you have a particular philosophy about marriage in light of your experience?

JPG: I think that it is very often the case that a man who fails in everything else, does not fail in his marriage.

SDC: It must have been an expensive occupation for you, too, getting married and divorced several times. I think the marriage service ought to be rewritten a little bit; it's not when you get married that you endow them with all your worldly goods, but when you get divorced, isn't it?

JPG: Yes, well put!

SDC: You've always been a man, as I understand, who has been very attractive to women. During the long separations due to business preoccupations, you were naturally inclined to look at other women. Did this cause jealous scenes at all?

JPG: I don't think so, except possibly with my first wife, Jeanette.

SDC: What happened there?

> **On marriage**
>
> Jeanette [his first wife] was of a jealous temperament and we got into arguments. I would say that some of the arguments were not my fault. For instance, shortly after we were married, we'd go into a restaurant and some pretty girl might smile at me. My wife would say: 'Who's that little prostitute ogling you?' And I'd say: 'That's not a prostitute, that's a very nice girl and an old friend of mine.' She'd reply: 'Well, you can't tell me she's a nice girl if she's ogling a married man.' I'd answer: 'Well, she probably doesn't know I'm married.' I had not been married very long, and there had been no publicity about it, and it started a row, you see. . . . I think I would have made a better boyfriend than husband.

JPG: Jeanette was of a jealous temperament, and we got into arguments. I would say that some of the arguments were not my fault. For instance, shortly after we were married, we'd go into a restaurant, and some pretty girl might smile at me. My wife would say: 'Who's that little prostitute ogling you?' And I'd say: 'That's not a prostitute, that's a very nice girl and an old friend of mine.' She'd reply: 'Well, you can't tell me she's a nice girl if she's ogling a married man.' I'd answer: 'She probably doesn't know I'm married. It wasn't in the newspapers; there hasn't been any publicity, so the chances are nine out of ten that she doesn't know I'm married. And even if she does, why shouldn't she smile at me? I've known her for years!' Well, it started a row, you see.

SDC: How long did that first marriage last?

JPG: We were married in October of 1923 and divorced in 1925.

SDC: Two years. Well, perhaps you were in a better position than most of us to settle these wives in a manner they felt they were accustomed to. Also, you've perhaps been able to terminate these marriages quicker than most people could afford to do. Did they all ask for enormous sums or were some kinder than others?

JPG: Judging from other divorce cases, they were fair.

SDC: Who was your second wife?

JPG: Allene Ashby. I married her in Mexico.

SDC: Presumably, the reason you married in every case was that you fell in love?

JPG: Yes.

SDC: That's the only reason, presumably, that would be sufficiently important to detach you from your business interests even long enough to get married. Did you meet her in Mexico?

JPG: No, I didn't. Frankly, I never got married with the idea that I was going to be an ideal husband because I think I was intelligent enough to know myself. I'd be considered a better boyfriend than husband. I was aware that I had a tendency to put business first. There's the theory that I had responsibilities in business which were optional with me, and nobody was insisting that I be president of an oil company; and if I wanted more time to myself, all I had to do was resign. But as long as I was president and hadn't resigned, I felt the company's business took priority. I realized that such an idea wasn't going to sit very well with the average woman. The average woman doesn't like to play second fiddle to business.

SDC: Do you think that most of them hoped to change you?

JPG: I don't think so.

SDC: They say that a woman, when she's thinking of wedding, thinks of three words: 'Aisle, Altar, Hymn'. And this is an approach to marriage that is not easy.

JPG: I think also, as my mother said, I made a mistake in marrying women who were so young. I think if I'd married a woman of my own age, one who would have had more experience in life, she might have been more indulgent, made more allowances.

SDC: Were they all much younger than you?

JPG: My first wife was eighteen when I married her, the second was about eighteen. And the third one was – let's see, Ann was about twenty-two – and Fini was about nineteen. The oldest was Teddy; she was twenty-four or twenty-five. So, they were all twenty-five or under.

SDC: So, you were obviously attracted to young women.

JPG: Well, I was younger then myself!

SDC: Yes, but even so, you suggest that there was a gap.

JPG: That was my mother's view, yes. You see, at the time of my last marriage – in 1939 – I was forty-six and my wife twenty-five.

SDC: Not a bad bracket in some ways. Did your mother get on with any of them at all well?

JPG: She got along with most of them fairly well. There were never any mother-in-law problems.

SDC: Tell me about wife number three.

JPG: That was Fini Helmle. She was German. Very good-looking. She's never remarried.

It was amusing how I met her. I was in Vienna, staying at the Bristol Hotel. I saw a man and his wife, obviously they were father and mother, and two girls at a nearby table. One of the girls was very beautiful, and I was hoping to meet her; however, I couldn't contrive anything. Finally, I sent a note to her, in which I invited her to dinner. I got a note back that she would meet me in the lounge for tea. So we met and got acquainted. She spoke very little English, and my German wasn't too wonderful. Then, we had dinner. I happened to pass her the next day and she looked at me with great hostility. I thought, what have I done! She had seemed charming and friendly the day before, and she was looking as if she could bite nails in two! I thought I might as well find out what the trouble was, so I asked her. She told me I wasn't a gentleman. 'What have I done?' I asked. I found out that, for some reason or another, the waiter had gone to her room in the hotel and handed the bill to her for the dinner rather than putting it on my hotel bill. So, she thought to herself that I was a bounder! Well, I apologized and said it was the fault of the hotel, that I never dreamed that they were supposed to pay for the dinner.

SDC: That wasn't a very good start; you were lucky to see her again at all. So the romance matured after that?

JPG: Yes.

SDC: How long did the marriage last?

JPG: We were married at the very end of 1928 and divorced about June 1932.

SDC: A short marriage. What happened there?

JPG: Let's see. We were married in Havana, Cuba, and we drove from Miami to Los Angeles and lived at my parent's home. Again, I had a flurry of business, and she was rather lonely because she didn't speak perfect English and didn't know a great many people. She wasn't a pushy sort of type who made friends easily. Then, she became pregnant, and she wanted to have the baby born in Germany. Her father was ill, and her mother had died very suddenly and unexpectedly from an embolism during a slight operation. That was really the problem. So, she went back to be with her father and stayed on to have the baby born. In fact, I went to Germany as soon as I could arrange it to be with her. I was with her at the time the baby was born.

However, I got an urgent cable about my father's health that April and left within the hour. There were no airlines in those days, so I went by boat. I was in Los Angeles nine days later. She wouldn't come with me; she wanted to stay in Europe. I couldn't live in Europe at that time; my business connections and everything were in California. That was what brought about the divorce.

SDC: Wasn't your fourth wife, Ann Rork, an American?

JPG: Yes. She was the daughter of Sam Rork, the movie producer. He brought out Clara Bow. She was the only daughter of Sam Rork – a very charming young lady. But again, I was submerged in business, and I think that women don't like to feel that they're an unimportant part of a man's life, and if he always puts business first.... I think a man is a business failure if he lets his family life interfere with his business record. We see that very often in our company. A man has every characteristic that would spell success, except he can't handle his family. He's submerged in family troubles, always gives the family priority. Of course, business is no place for him.

> **On marriage**
> I think that women don't like to feel that they're an unimportant part of a man's life, and if he always puts business first.... I think a man is a business failure that lets his family life interfere with his business record. We see that very often in our company. A man has every characteristic that would spell success, except that he can't handle his family. He's submerged in family troubles, always gives the family priority. Of course, business is no place for him.

SDC: That's a harsh doctrine for married men but one that you've no doubt found applicable.

JPG: Yes. If you were going to buy shares or stock in a company, you'd rather have a president who always gives priority to business rather than to his family troubles.

SDC: So, you think that again was the cause of the break-up with Ann?

JPG: Yes. We just gradually drifted apart.

SDC: I'm interested in the dramatic circumstances when your fifth wife, 'Teddy' Lynch, was arrested immediately after your wedding as an American spy while you had to return to America from Rome.

JPG: We were married – let me see, when was it – November. The war was on but Italy was neutral then. She was working for an American newspaper in Rome as the Rome correspondent. After Italy went to war – probably in June 1940 – she and many others were arrested. They were sent to Siena. They were reasonably looked after, but they couldn't leave the city. They finally got a Swedish steamer, the *Gripsholm*, and it brought home a boatload of Americans. She came immediately to Tulsa, where I had a war job.

It was the first time we had seen each other since the day we were married, because we were married, then I had to go down by train that night in order to catch a boat in the morning.

SDC: It must have been a very shattering experience for you both. You had to separate for the best part of two and a half years!

JPG: Yes, it was.

SDC: And you couldn't see each other again. Did this, in itself, undermine the future of the marriage, do you think? I mean, the fact that you had to start under this separation?

JPG: I suppose it did, looking back.

SDC: Had you known her for some time?

JPG: Yes, some time before we married. I met her in the summer of 1935 in New York.

SDC: When you were able to join up in Tulsa, you were running Spartan Aircraft, weren't you?

JPG: Yes, and I often wonder about my performance there. It was a big war plant, and of course, conditions were very difficult. We had a lot of employees who didn't know one end of the factory from the other. Plus, the moment you got good men, they were drafted and you lost them. I was working fifteen, sixteen hours a day, and we were going twenty-four hours a day. I tried to be there every shift. Consequently, I didn't have much time for my marriage after she joined me.

SDC: Were you aware at the time that this was eroding her confidence in the marriage? Although a lot of wives might have expected that, during the war, preoccupations with an aircraft factory would be enough to justify a very considerable absence. After all, many of us had to leave our wives for a year or more during the war.

JPG: Yes. Well, she had a singing career, and like many women, she was an extremist – very interested in the singing career. She hadn't been with me in Tulsa too long before she got an offer she felt she couldn't overlook to sing in California. So, she went to California and I was tied to my job in the aircraft factory. So, we didn't have much time together even after she got back to the United States. She came from a good family, a good social family. She was in the Social Register. But there was a fad, and she had a very good voice. She was taking singing lessons and then singing professionally. She was very intense about her career. I think that makes difficulty in a marriage, too.

SDC: She was in fact a career woman?

JPG: Yes. I don't know whether you saw *Lost Weekend* with Ray Milland. She was the one who sang the opera part – a very pretty woman.

SDC: How long did it last?

JPG: She came back in 1946, and we separated in 1951. We didn't actually get a final divorce until 1958.

SDC: Was that your last marriage?

JPG: Yes.

SDC: Looking back on it now from your rather impregnable position here in Sutton, do you feel on the whole that life has been a little easier and quieter as a bachelor man?

> **On marriage**
>
> Marriage is a responsibility. You have to work at it. It's like a garden, I suppose. The more care and attention you give to a garden, the better off it is. Since I've been here, I've had a lot of business to look after. I suppose there are advantages and disadvantages to being married . . . . The disadvantage is that you hate to do a poor job. And then, I have a bad trait, I suppose, from a woman's standpoint, and that is I have been brought up that business comes first. I stay late at the office and bring a suitcase full of homework. By the time I get through the homework, the wife is long since asleep!

JPG: Marriage is a responsibility. You have to work at it. It's like a garden, I suppose. The more care and attention you give to a garden, the better off it is. Since I've been here, I've had a lot of business to look after. I suppose there are advantages and disadvantages to being married. I think in my position, with public shareholders and people working for the company, I like to settle the problems each day; I don't like to postpone things. Obviously, that makes some fraught days for married life . . . . I stay late at the office and bring a suitcase full of homework. By the time I get through the homework, the wife is long since asleep!

SDC: Have you found that these divorces have been very searing periods, very temperamental situations?

JPG: I think it's a great strain. As you know yourself.

SDC: I do, indeed.

> **On lawyers**
> You can imagine a woman going to a lawyer, and getting an interview, and announcing that her name is Hughes, Mrs Howard Hughes. The lawyer rubs his hands and says: 'You're Mrs Howard Hughes, *the* Howard Hughes?' She says: 'No, the lifeguard down at the beach, but he has the same name.' That lawyer is going to lose interest right away!

JPG: You always feel that you're at a disadvantage if you've got any position and any property because you might get unfavourable publicity, and you have to pay through the nose. You can imagine a woman going to a lawyer, getting an interview, and announcing that her name is Hughes, Mrs Howard Hughes. The lawyer rubs his hands and says: 'You're Mrs Howard Hughes, *the* Howard Hughes?' She says: 'No, the lifeguard down at the beach, but he has the same name.' That lawyer is going to lose interest right away!

SDC: Yes, of course. I knew one man who'd already had two wives, and they'd each been given a third of his income. When he had his third divorce, the judge said that his wife was to have the remaining third of his fortune and income. And he said: 'Well, in that case, I'd sooner go to prison than pay any of you. I won't pay a penny to any of you because if you're going to take three-thirds of my income, there's not much left for me. So, what you can each do is come and spend four months a year with me, in rotation, and look after me.' And he persuaded them to do it. So, he managed to get his third back from each of them in four-month co-operation.

JPG: Good bargain.

SDC: Looking back on it all, do you feel you've made any mistakes?

JPG: I think you always wonder whether if you'd married

other women you've known well, you might not have been better off. I suppose a woman thinks that herself. She looks back at beaux she has had and wonders if she hadn't married you or me – if she had married Charlie – she might be better off.

SDC: Who was the Duchess of Carcaci? Do you have a photograph of her?

JPG: She was the daughter of Effie and Eugene Millington-Drake. He was the famous ambassador in Uruguay during the war. Her mother was the daughter of the original head of the P & O, Lord Inchcape.

SDC: What was her Christian name?

JPG: Marie. She married a Sicilian. Like the Leopard.

SDC: Was he the origin of the Leopard?

JPG: No.

SDC: Had you known her for a long time?

JPG: I met her in 1949. She was very beautiful. Lawrence Durrell wrote about her in *Bitter Lemons* [1957].

SDC: She was a personal friend? A person you knew well? You have other great women friends?

JPG: Penelope Kitson was very influential in changing my mind over buying a place in England rather than in France. We were great friends.

SDC: You gave a dance for her daughter. Have you seen much of her lately?

JPG: I see her occasionally. Jeannette Constable-Maxwell is another good friend of mine. Her father's a very good friend of mine, too. I was a little embarrassed when, in the summer of 1959, they invited me to stay with them at Farlie House in Scotland. The invitation was for myself. I arrived and then business associates began to arrive. Farlie wasn't close to any hotel, and of course, the Maxwells are very hospitable. So, I shortly found myself surrounded by four or five businessmen. I felt embarrassed about it because they had invited me there, and I had a small army.

SDC: There have obviously been women in your life who have had a great influence on you and to whom you've been very attached, apart from your wives. Who would be the one . . . ?

JPG: I don't think any woman ever had much influence on me in business. I think that the only business decision I made which was partly influenced by women friends was the purchase of Sutton Place. Penelope Kitson was very instrumental. So was Alice Clifford, the wife of Bede Clifford who lives nearby and was a great friend of Geordie. Mary Teissier was, also.

SDC: Was Mary Teissier, at that time, interested in your being in England in Sutton?

> **On early romances**
> I was travelling in 1913 and had a rather romantic meeting in Constantinople. I met a woman there who was a few years older than I was. I was twenty at the time. She might have been twenty-five. She was unhappily married to a man who was the Russian Consul-General in Asia Minor. . . . She was French. Her name was Marguerite Tallasou. We saw each other a few times in Constantinople, and then she left. She had to go to Brusa. I said goodbye to her at the Mudania Pier in Istanbul, and she was crying. I think that, with the instinct women have, she saw it was the final farewell. We wrote for about a year, until the war broke out. Then, of course, it was impossible to get any more letters. I never heard from her again. What's happened to her, whether or not she's still alive, who knows?

JPG: Not particularly, but she thought that Sutton Place was very attractive, so I found myself buying a place in England. Of course, the average American is more inclined to live in Paris. London is supposed to be solid and substantial, Paris gayer. I used to come over on the boat from New York, and there were about 600 people in the first class. The boat would call first at Cherbourg or Le Havre and then go on to Southampton. About 570 people would disembark at Cherbourg or Le Havre and go on to Paris, and there would be only thirty left to go on to England. Most of them felt it was incumbent upon themselves to explain why they weren't going to Paris. They had relatives in England or had important business, and it was considered essential to explain why you were going on to England.

SDC: Any other great women friends?

JPG: I was travelling in 1913 and had a rather romantic meeting in Constantinople. I met a woman there who was a few years older than I was. I was twenty at the time. She might have been twenty-five. She was unhappily married to a man who was the Russian Consul-General in Asia Minor.

SDC: Was she Russian, too?

JPG: She was French. Her name was Marguerite Tallasou. We saw each other a few times in Constantinople, and then she left. She had to go to Brusa. I said goodbye to her at the Mudania Pier in Istanbul, and she was crying. I think that, with the instinct women have, she saw it was the final farewell. We wrote for about a year, until the war broke out. Then, of course, it was impossible to get any more letters. I never heard from her again. What's happened to her, whether or not she's still alive, who knows?

SDC: You never had any idea what happened to her during the Revolution?

JPG: No, no, I never did. So the mists of time have intervened. (*Getty did not attempt to hide his tears.*)

SDC: Yes. But from what you say, she was a person who touched you deeply.

JPG: Yes.

SDC: And you must have felt very sad at being severed like that, unable to make any contact.

JPG: Yes, yes. (*The words seemed to be wrenched out of him by the haunting recollection.*)

SDC: Did this haunt you for quite a time?

JPG: It did in a way, but of course, I had many interests. I didn't dwell on it, but it was regrettable that I never saw her again.

I felt somewhat embarrassed at prying into this secret of his past and changed the subject abruptly.

# 6
# FAMOUS FRIENDS

John Paul Getty was a generally quiet, unobtrusive man who nevertheless filled his life with prominent and fascinating friends and acquaintances. Here, we talk about just a few.

SDC: Let's talk about your time at Oxford for a bit. You went there in 1912, I think.

JPG: Yes, I had a small apartment at 12 High Street, very close to Magdalen. I was a non-collegiate scholar there; I didn't belong to any particular college. I was just taking economics and political science courses. But I was very close to Magdalen because all the lodgers at High were Magdalen men.

SDC: Is that where you met the Prince of Wales?

JPG: Yes. He was a good friend of a man called Alistair Menzies, and he'd visit him once or twice a week. I was quite intrigued because he had what was called a bulldog, a personal detective. If he whistled, this bulldog or detective came running.

SDC: How did you get to know him?

JPG: I was introduced by one of the lodgers. That was the start of a long friendship. I saw him at Oxford quite often.

SDC: What were his interests at that time?

JPG: I don't think he was serious, certainly not in university precincts!

SDC: Did he do any work at all?

JPG: I suppose he must have done some!

SDC: Did you have a few romances at Oxford?

JPG: Yes. There was one girl I took out a few times . . . a friend of the Prince of Wales. He'd taken her out. I don't think either of us was very serious about it.

SDC: When did you next see the Prince?

JPG: The next time I came across him was in 1930, actually I believe it was 1934, in Biarritz. I just said 'hello' to him and exchanged a few words. He was very cordial and asked me to look him up. But I left Biarritz the next day. The next time I saw him, he was the Duke of Windsor, in Paris. As a matter of fact, I saw quite a bit of him. He was going to invest some money on the New York Stock Exchange, and he asked me what I would recommend. I told him Mission Corporation, so he bought some shares, and right afterwards the market had a sinking spell and Mission Corporation went down. He had quite a paper loss. I used to dread meeting him.

> **On the Duke of Windsor**
>
> The next time I saw him he was the Duke of Windsor, in Paris. As a matter of fact, I saw quite a bit of him. He was going to invest some money on the New York Stock Exchange, and he asked me what I would recommend. I told him Mission Corporation, so he bought some shares, and right afterwards the market had a sinking spell and Mission Corporation went down. He had quite a paper loss. I used to dread meeting him. Ultimately it went up, and he had a very nice profit. He sold out at a good profit so I didn't have to run down an alley if I saw him coming.
>
> He was very good company, great charm. He was a very intelligent man too, in his way, in sizing up people and political trends. It had been his education. I've often thought that the full truth has not come out regarding his abdication. I sat up with him until two or three o'clock in the morning sometimes, discussing old times, politics, and so forth. While he didn't say so, I came to the conclusion that there are some people who will sign anything that is set before them, and other people won't. They'll want to study it, then revise it a little bit. He belonged to the latter school.

SDC: It was your own company, Mission Corporation?

JPG: Yes. Ultimately it went up, and he had a very nice profit. He sold out at a good profit, so I didn't have to run down an alley if I saw him coming.

SDC: Was he good company?

JPG: Very good company, great charm. He was a very intelligent man too, in his way, in sizing up people and political trends. It had been his education. I've often thought that the full truth has not come out regarding his abdication. I sat up with him until two or three o'clock in the morning sometimes, discussing old times, politics, and so forth. While he didn't say so, I came to the conclusion that there are some people who will sign anything that is set before them, and other people won't. They'll want to study it, then revise it a little bit. He belonged to the latter school.

As I see the English monarchy, and I may be wrong, but I think the monarch has very great influence but practically no authority whatsoever. The monarch is supposed to sign everything that is submitted to her; I think that's particularly true nowadays. It was also true of the last George V and George VI. I think Edward VII had a certain discretion about what he signed, and I know Queen Victoria did. For instance, when they had the note for her signature about the South in the Civil War, she turned it over to Albert. He read it and said: 'That could involve us in the war.' He said: 'Change it,' and so she changed it.

SDC: Things have moved on a bit since then. I was in the House of Commons during the abdication, and everything stampeded at such a pace that it was difficult for anybody to form a detached opinion. I do remember the atmosphere of the public was very strong. The feeling against Mrs Simpson was very strong. When I went to my constituency in south-west Norfolk that weekend, people were grabbing me on the street, trying to influence me to get rid of her.

JPG: He was quite popular, wasn't he?

SDC: He had been immensely popular until that moment, but the extraordinary radical and puritanical streak in the British people came out very strongly against him on this occasion. Winston Churchill was in the House at that time and was defending him – not only as a personal friend but on the grounds of constitutional propriety. He was saying that you cannot monkey around with primogeniture in matters of the Crown because if you decide in the House of Commons that you don't like number one because of his wife, you might not like number two because of his politics, and you don't know where you're going to finish. You are setting yourselves up as an electoral college, and in the end the monarchy would suffer from this. You have to take your first-born heir-apparent as he is. That is the way the hereditary system has worked. It has its disadvantages, but it also has its advantages of continuity.

JPG: Yes, yes. (*Getty agreed enthusiastically.*)

SDC: Although we haven't suffered in the short run because we have a very good Queen and we had a very good King and Queen Mother, in the long run it could be said that the succession to the throne has not been altered in any way – the Duke of Windsor had no children – the present Queen would have come to the throne on his death.

JPG: I knew Ernest Simpson, too.

SDC: I met him. He was a young officer in the Guards.

JPG: Yes. He was very pleasant.

SDC: He never talked about the episode at all.

JPG: No, he never did. He was very much a gentleman about that.

SDC: Did you know Mrs Simpson, too?

JPG: Oh yes, I knew her. (*Getty leaned back in his chair to tell the story with evident relish.*)

You know I was a good friend of the Furnesses, Thelma Furness. The story is that Thelma Furness was the girlfriend of the Prince, but she had to go to New York. So, she searched the list of her acquaintances for a girl she thought would be perfectly safe for the Prince to know. She settled on Mrs Simpson as charming, but not too charming, and seductive, but not too seductive. She had been gone some weeks and when she came back, she'd lost her pride of priority. She wasn't the number one girl anymore.

SDC: No. Mrs Simpson got that position and retained it with great authority.

JPG: Yes. Mrs Simpson had great charm and great efficiency, too. She ran the house to perfection, and she was a good manager. She had the gift for society. I don't think, however, that the Prince was obstinate about trying to force Mrs Simpson on the British public. I don't think that he expected to force a marriage either. I think that events moved so fast that he found that he was programmed to abdicate before he really knew what was happening. I think that Baldwin engineered it.

SDC: He did. He was a very crafty man, Baldwin.

JPG: And the editor of *The Times* at that period [Geoffrey Dawson]. I think that the reason they wanted to get rid of the Prince was not too much due to Mrs Simpson, but they felt he was too much to the Left politically to be a great candidate. I think if he'd had different political views – it's just my opinion – but I think he would have settled for some sort of marriage of convenience, a morganatic marriage. Because it was all about what never happened, anyhow. She didn't have a child.

SDC: Would she have settled for a morganatic marriage, do you think?

> **On the abdication**
> I think that the reason they wanted to get rid of the Prince was not too much due to Mrs Simpson, but they felt he was too much to the Left politically to be a great candidate. I think if he'd had different political views – it's just my opinion – but I think he would have settled for some sort of marriage of convenience, a morganatic marriage. Because it was all about what never happened, anyhow. She didn't have a child, never was to have a child. As I picture her, what I knew of her, I don't think that she wanted him to do anything that jeopardized his position. She realized he wasn't just an ordinary man walking down the street. He was the head of the Church and wouldn't be the head of the Church. He had political responsibilities. . . . I don't think that she tried to high-pressure him into marriage.

JPG: I think so. As I picture her, what I knew of her, I don't think she wanted him to do anything that jeopardized his position. She realized he wasn't just an ordinary man walking down the street. He was the head of the Church and wouldn't be the head of the Church. He had political responsibilities. The fact was that she was a divorced woman, I think she was sensible and realized that. I don't think that she tried to high-pressure him into marriage.

SDC: At the time, the Archbishop of Canterbury was very much against his marrying her because of her divorce, and the Church was very influential then, much more so than it is now.

JPG: Well, looking back on it, from today's viewpoint, if you're going to ostracize everyone who's had a divorce . . . (*Getty laughed.*)

SDC: Yes, I think things have changed a bit. Do you know any of the present Royal Family?

JPG: I know the Duke of Edinburgh, yes.

SDC: How do you get on with him?

JPG: He's a very charming man, very charming. I was much touched when I was with him at a reception a few years ago. We were going down a long flight of steps and when he got to the bottom, he opened the door and held it open, motioning me to go through because I was the older man.

SDC: That was rather nice. He's got a very good brain, hasn't he?

JPG: Very, very smart. I think if he weren't of royal blood, he could be a good politician. He might have been a Prime Minister. I think he'd be very effective at electioneering.

SDC: Do you know the Queen, too?

JPG: Yes, she's also very charming.

SDC: I see you've got a picture of the Queen Mother there on the table.

JPG: Yes, she came here for dinner.

SDC: Obviously, you've got to know a great many of our leading personalities in England. This doesn't arise only from your period here at Sutton, does it? You've known them over a longer period than that? Are there any other people in England you particularly like to see much of?

JPG: Well, I'm very fond of the Bedfords. I give him a lot of credit because I don't think he ever expected to make any money out of Woburn, his stately home. I think he did it because he felt that it was his responsibility to keep the place going, and he's worked hard to do that. It is very successful. I've been a guest several times at Woburn, and people tell me that it's better run now than in the time of his grandfather fifty years ago!

SDC: Of course, the price he has to pay for that is to have the place pretty well swamped by the public in vast numbers!

JPG: Yes, I was with a French friend of mine at Woburn for a weekend party some years ago, and he was told that 5000 people visited the house. He made the remark that in France if 5000 people came to visit your home, they came to loot it. (*Getty chuckled.*)

SDC: Who were some of the interesting people you've met in your travels?

JPG: I met Prince Yusupov in 1935 when I returned to London from a trip to Russia. I ran into Prince Yusupov, and he invited me down to Windsor. The Russian royal family at that time had been given a house in Windsor Park by the king. I had a lot of photographs taken of Yusupov's house in Leningrad, and they were very much interested in these photographs. The whole royal family was there.

SDC: Did you ever talk to him much about his murdering Rasputin?

JPG: Yes.

SDC: What did he say about that?

JPG: He said Rasputin was a spy and that's why he killed him.

SDC: It's never been exactly proved that he was a German spy – just that he had this tremendous ascendency over the Tsarina and was influencing her to make peace with Germany. Anyhow, Yusupov would then have been a man in what time of life when you knew him – about sixty?

JPG: He was eighty when he died, so he was in his early fifties, I suppose. Very young for his age.

SDC: I had a letter from him. I wanted to know if he thought it was possible that any of the royal family had survived the holocaust at Ekaterinberg. I studied Russian with the nuns of the Russian Orthodox Church at Gethsemane outside Jerusalem when I was convalescing from wounds during the war. There were a lot of Russian nuns there, and they were talking a lot about a Grand Duchess

Elizabeth who had come out there to Jerusalem before the Revolution. There was a Russian nun there called Sister Varvara who could have been with the Archduchess at the time.

JPG: Yusupov told me that this woman who said she was the Grand Duchess Anastasia spoke Russian with a strong Polish accent.

SDC: He didn't think she had any claims to it?

JPG: No, he didn't. He said she had a Polish accent.

SDC: She could have acquired the accent in some way. She was fairly young at the time of the assassination; and if she got out at all, she was supposed to have travelled for nearly two years with peasants across Russia, and they might have been Polish.

JPG: They might have been, yes.

SDC: Did you ever see more of Yusupov after that?

JPG: Yes. I saw him several times. He was quite a friend of mine, as was Mary Teissier. She was the great-granddaughter of Alexander II. She was a Romanov and related by marriage to Yusupov. He was a very fascinating man.

SDC: In what way?

JPG: Well, he played the banjo very well and would sing Russian songs. Very charming, and one of the handsomest men I've ever seen.

SDC: Did he spend any length of time in England? I didn't realize he'd been in England long.

JPG: No, he lived in Paris. He told me that when he left Russia, he took out two pictures by Rembrandt. They had thirty-five pictures by Rembrandt in the gallery in Russia, and he got two he took out with him. He tried to get the jewels out but couldn't. He lived on the proceeds of the two Rembrandts. Only they found that one of them wasn't a Rembrandt. However, it was worth just as much – it was a Vermeer!

SDC: Yes, of course that would have been just as nice. Which was the Rembrandt, do you remember?

JPG: I don't remember, but they were good-sized paintings. And he made a loss on them because he borrowed money from Weidner on the paintings, and then he sold them again to pay Weidner back. Then Weidner laid claim to the pictures, so they had a loss. I think they're in the National Gallery now. He also had a case with Metro-Goldwyn-Mayer over Rasputin . . . they did a movie of Rasputin. But actually, he was popular with the Soviet government.

SDC (*astonished*): Was he?

JPG: If he had wanted to go back to Russia, they would have welcomed him.

SDC: Why was that?

JPG: Because he killed Rasputin.

SDC: That's interesting.

JPG: Yes, it was . . . as far as the Soviets were concerned.

SDC: But he would have had to live a very different life in Russia to what he was accustomed to before. I mean, he wouldn't have found it very easy to settle down in the Soviet Union, I think. So, he remained in Paris and that's where you saw him most, I suppose?

JPG: Yes, in Paris. It was amusing. I knew he liked caviar, and a friend of mine was invited to lunch by Yusupov and took him a large jar of caviar. Diana Cooper was at lunch, and the prince gave the caviar to her. He was very gallant. He liked it very much himself, but there was a lady present, and he handed it to her. [Diana Cooper later divulged to me that she hated caviar!]

SDC: That was very nice. I had some caviar at a restaurant on the way up to London the other day after I'd been talking to you. For about two teaspoonsful, they charged me, I think, about £2.30.

JPG: Yes, it's terrible.

SDC: The whole thing's ridiculous, the price of things.

JPG: When I gave my dance here in June of 1960, I had about 70 pounds [32 kilograms] of caviar for the supper and I paid, I think, something like £12 a pound [0.5 kilogram] for it.

SDC: You've given marvellous parties here; it's a wonderful setting for them. Do you enjoy entertaining?

JPG: Yes.

Sutton Place is perfect for parties with its Great Hall – 51 feet (15.5 metres) long, 25 feet (7.5 metres) wide, 31 feet (9.5 metres) high – where you can find the carved arms, coronets and garters of the Bourchiers, Fitzalans, Howards, Paulets, Fitzwilliams, Russells and more famous families. The Long Gallery was where Queen Elizabeth I greeted subjects while visiting her cousin Dorothy Arundell Weston. In fact, Sutton's seventy-two rooms included more than a dozen reception rooms.

Paul may have needed all of them at the first huge party he gave at Sutton. Twelve hundred people were invited; twenty-five hundred showed up. He later remarked: 'It seemed more like twenty-five thousand to me. When I was finished with the receiving line, I was finished!'

# 7
# EARLY TRAVELS

Much of this tape of Paul's travels was recorded on 8 February 1974. It begins with his first trip in 1912.

SDC: Paul, the book is getting somewhere near completion, but there are certain loose ends I want to try and tie up with your help. You've referred to some earlier travelling. Can you give me some more information?

JPG: Of course. I went to China and Japan in 1912. I left San Francisco on the *Shinyo Maru* of the Toya Keisa Kishu Line in May and went to Honolulu. We stopped a day there, I remember, and I enjoyed a swim at Waikiki Beach. I went to the hotels, saw the island, went back on board, and left in the evening. It's a very colourful ceremony when you arrive in Honolulu; the women come out singing songs and putting leis [garlands] around you. Then, we went to Yokohama. I remember one incident: there was a young man who got on board with an automobile and played cards every night. About three nights later, I heard he'd lost the car.

SDC: He gambled it away! A car was a very valuable commodity then!

JPG: Yes. Anyhow, we landed at Yokohama, but I saw little of it. An amusing incident happened in the harbour, though. The ship's doctor was a friend of mine, or supposed to be a friend. I put on a bathing suit and was going to take a swim. He failed to alert me of unforeseen circumstances, so I dove off the ship's ladder and went into iced water. I came up out of the water in a second, and he was roaring with laughter!

SDC: Why, because it was so cold?

JPG: It seems that the ship was discharging refrigerated water, and I dived into it. A terrible shock – real iced water.

SDC: So, he had his tongue in his cheek all the time.

JPG: Yes. Then, we went on to Tokyo and out to Miyanoshita. We went through the inland passage, past Fujiyama, down to Kobe. The ship's captain was an Englishman. I think they had some trouble with insurance, and they had to have an English captain at that time. His wife lived in Kobe, and he had one half-hour with his wife every two months. If she'd lived in Yokohama or Tokyo, they'd have had more time.

SDC: Cutting it rather fine.

JPG: Perhaps that was all they could stand of each other! Then, we went to Nagasaki.

SDC: What was it like in those days – before it was bombed?

JPG: A beautiful city. Lovely, the old city. Then, it was Shanghai for a couple of days. I was much impressed with Shanghai; I liked it. Then, it was Hong Kong.

SDC: So, you really only called on the mainland of China for a couple of days. You never got to Peking?

JPG: No, I never did. However, I went up to Canton.

SDC: Did you find the atmosphere very friendly there in those days?

JPG: Well, there was a plague when I was there. I wore a mask. They had pneumonic plague, and some of them were dying. They would be perfectly healthy and three hours later, they would be dead.

SDC: So, you were exposed to some risk going to Canton.

JPG: Yes, I didn't see much of Canton! Then, I got a small boat – the *Prince Waldemar*, of the Nord German Lloyd line – and went to Manila. We slept out on deck because it was very hot and rough. Every once in a while we'd get a splash of salt water while we were sleeping.

SDC: Was that the only time you've been to Manila?

JPG: The only time. They had a place spelt B-A-Y (pronounced 'buy') outside Manila. It's a resort area, and there was a big bar in the Manila Hotel. There was quite a game there, as I found out later. They would ask a newcomer, a tenderfoot, where he was going. He might reply: 'I'm going to Bay,' because that was a great target for tourists. So, the moment he uttered the fateful words: 'I'm going to Bay [buy],' everybody lined up at the bar, and the tourist had to pay for all the drinks. I got caught. I remember there was an Englishman there, telling me about the wisdom of investing in copra.

SDC: The sort of stuff made from coconuts?

JPG: Coconut oil. He said that if one could put £5000 into copra, in five years one would have an income of £5000 a year. It was a lot of money in those days. So, I said: 'Why didn't you do it?' He said: 'Well, I never thought I'd be here long enough.' I asked: 'How long have you been here?' He answered: 'Twenty years.'

SDC: So you went into another kind of oil.

JPG: Yes.

SDC: I'm not surprised. You seem to have been getting around the whole world at this time.

JPG: I went from Manila back to Hong Kong, to Shanghai, Japan, Honolulu, and finally, San Francisco.

SDC: Did you like Japan very much from a scenic point of view?

JPG: Yes. It was a beautiful country then, beautiful and very quaint. They weren't wearing Western clothes. I've never been back since . . . nor China. Of course, when I was in China, it was only ten years after the Boxer War.

SDC: Were the effects of it apparent in any way? Were the sentiments of the Chinese you met still very hostile to Westerners?

JPG: Yes, there was Young China. The Young China Movement, politically. This would have been just after Dr Sun Yat Sen became President. I think the Emperor was just being deposed.

SDC: Did old China seem prosperous at that time?

JPG: I was much impressed with China and the Chinese – very hard-working, intelligent people. I felt that China would play a great role in the world.

SDC: As Napoleon said: 'A giant is sleeping, do not wake him.' Do you remember?

JPG: Yes, tremendous power in China. I felt that in Japan, too.

SDC: Now that you have seen this tremendous development of China since then, what part do you feel it is going to play in the world?

JPG: Well, the time may come, whether we like it or not, when China is the most powerful country in the world.

SDC: You visited another powerful country in 1913, didn't you? Russia?

JPG: Yes. As soon as I graduated from Oxford with a diploma in economics and political science, I left on a long tour. First to Germany – Berlin – then Denmark, Sweden, Abo [now Turku], Finland, then Helsingfors; St Petersburg [now Leningrad], Russia, and Moscow, Nijni Novgorod, then to the Volga River and the Caspian Sea.

SDC: Tell me a bit more about your time in St Petersburg.

JPG: Well, I arrived in July 1913 and spent about two weeks there. I liked it very much and had a Russian tutor an hour a day. I learned to ask for milk.

SDC: *Moloquo.*

JPG: A glass of milk – *stakan moloqua* – and how to go to the railway station and ask for first class. I also learned the words: 'I am an American; I don't speak Russian.' I found that many people in Russia were unacquainted with foreigners. They'd come up and speak to me in fluent Russian and then couldn't understand my reply. If I didn't say anything, they'd think I was rude. Even when I spoke English, they didn't recognize it as a language.

SDC: Did you keep it up after that?

JPG: I have a Russian friend, Mary Teissiér, who gives me a little Russian practice.

SDC: What more do you remember?

JPG: The magnificent buildings. I went out to Peterhof and the various palaces – the Alexandra Palace. I own the tapestries that used to be in the Pavlosk Palace.

SDC: Did you actually see them at that time?

JPG: Yes. The Soviets sold them after the war. They were bought by Duveen and Company. When Duveen sold out to

Paul Getty in 1938 by Gerald Brockhurst RA (J. P. Getty Museum). He was forty-five; a more sophisticated man than the boy who had started on the oil rigs of Oklahoma.

Graduating class for the year 1906 from Emerson School, Minneapolis, Minnesota. 'The boy most likely to succeed' is already the centre of the picture.

Home of Getty's parents: Mr and Mrs George Franklin Getty – 647 South Kingsley Drive, Los Angeles, California – built in 1908.

The Monroe, Washington, home of Laura Getty Allen, Paul's aunt. Paul's mother wrote: 'Georgie and I had a delightful visit here. Such a wonderful family.'

Paul Getty in 1938–9; bust by Vignelli. Getty was touring Europe; and he was concerned with seeing beautiful pieces of furniture or carpets, noting the prices, and in many cases, making offers that were accepted.

George Franklin Getty, Getty's father, born on 17 October 1855.

Penelope Kitson accompanied Getty in Saudi Arabia; she was a great friend and was influential in his buying a house in England.

Paul Getty and the King of Saudi Arabia during oil explorations around 1953. Getty was to refer to the prophets of doom on Wall Street who thought he had met his Waterloo on the sands of the Arabian deserts.

Teddy Lynch, Getty's fifth wife, with their son, Timothy, who died under the anaesthetic during a minor facial operation.

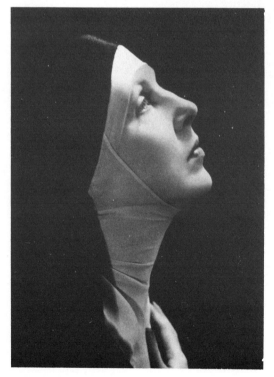

Lady Diana Cooper. She divulged to Getty: 'I hate caviar'.

Paul Getty receiving Queen Elizabeth, the Queen Mother, at Sutton Place.

Portrait of Somerset de Chair by Theo Platt.

Paul Getty at his desk in Sutton Place, built in 1521.

Norton Simon some years ago, he acquired them. Norton Simon sold them at public auction in New York about two or three years ago, and I bought them. The Russians came to me afterwards and wanted to buy them from me or trade me items for them. I said: 'Why didn't you appear at the auction if you were interested in them and buy them then?' They said: 'Well, we didn't know they were being sold; it hadn't been drawn to our attention.' I thought, well, the Russians supposedly know every time a pin drops in the United States.

SDC: Obviously not.

JPG: I kept them. They're at Malibu.

SDC: Can you describe some of your activities on the Caspian? This is all very interesting.

JPG: Yes, it was quite interesting. On the boat going down there, nobody spoke English but me. There was a Russian colonel on board who knew the following words: 'I love you, kiss me quick.' Anyhow, when we got down to Astrakhan, I was surprised when a steamer was not there to take us to Baku. I found out that we had to go 50 or 60 miles [80 or 100 kilometres] out in the Caspian Sea in a small boat before the water was deep enough for the bigger boat. When we got to the small boat, I found I had a cabin with a Russian prince! We had a very big storm during the night, and he was very seasick. I didn't get sick and was able to help him, and he seemed very grateful. Instead of going on by boat to Baku, we decided we'd had enough of small boats in the Caspian Sea, so we took the train down from Petrovski.

SDC: Were there any oil wells at Baku in those days?

JPG: Oh yes, there was great oil at Baku.

My father's monthly allowance was due, and I went twice a day to the Russo-Asiatski Bank in Baku, but I didn't get any money. Meanwhile, the Russian prince had got tired of Baku and wanted to go to Tiflis [now Tbilisi] in Georgia. I told him I had to wait, that I hadn't money to leave. He offered to lend me the money. He insisted, and I didn't want to go alone when the time came so I finally borrowed the money and left a forwarding address at the Russo-Asiatski Bank in Tiflis.

It was rather wild country there in those days. I was carrying a revolver; it had been recommended. On the train to Tiflis, a rather wild-looking man got into the compartment at the first stop and sat opposite us. Then, he pulled out a knife and started sharpening it. I thought it was about time to show I was equipped, so I pulled out the revolver and spun the chamber.

SDC: In the best Wild West tradition!

JPG: Spun the chamber a couple of times and kept it on my lap with my hand close to it. So the man got out of the compartment and was seen no more.

SDC: Very good, and then what happened? Did you stay with this prince?

JPG: Yes. I was very keen on climbing Mount Ararat, but the prince wasn't enthusiastic about it, and so I gave it up. I was still waiting for money, going to the bank twice a day, but the money didn't arrive. I thought I'd rather wait at Tiflis and send the prince his money wherever he was when it finally arrived. He thought it was better to go on as we had been, so he lent me some more money, and we went on. We hired a car and went over the Georgian military road and saw marvellous scenery – the great peaks in the Caucasus, including Mount Kafkaz, the highest peak at 18,000 feet [5500 metres]. He knew a prince who lived right at the foot of Mount Kafkaz, and the prince had lunch for us in the open air right at the foot of the mountains. There were fifteen or twenty servants, champagne, caviar, and Russian dancers afterwards.

We went to the Black Sea to a place near Batum. It was the first time I'd seen the Russian custom in those days of women bathing without wearing bathing suits. They bathed separately from the men. It was considered rather affectatious to wear bathing suits. So, I was riding with my friend along the road by the beach, and I suddenly saw about five or six girls without any clothes walking along the beach. I shouted: 'Look at those girls; they have no clothes on!' The prince looked over and said: 'Well, what about it? They're bathing.' He didn't see anything to get excited about. So, I was rather ashamed of my excitement, but one has to realize that there are other customs in other countries.

Then we took a boat and went to Kerch. A sneak thief went through our pockets on the boat, so the money I had, which had been borrowed from the prince, was taken. Fortunately, he'd kept his principal sum in his coat, and the thief had only gotten our trouser pockets. I'd been in China shortly before and had been carrying some Chinese and Japanese coins, which I lost to the thief. The prince was very annoyed at being pilfered and called the Chief of Police in Kerch to board the boat, and in many gestures described this great loss we'd suffered, including a collection of 'Yaponski' and 'Kitaski' coins. He made it sound as if I'd had one of the great coin collections!

SDC: Did they catch the thief?

JPG: No, we never saw the money again. We went on to Yalta, but it wasn't the season there; however, I enjoyed it very much. Then, we went on to Poland. I was going to Vienna, and the prince was going to Rome. I said: 'Why don't you come to Vienna for a few days, and then I'll join you in Rome?' He said: 'Me, go to Vienna? I'm an officer in the Tsar's bodyguard. They take a very poor view of us in Vienna. They'd wonder what I was doing there; I'd probably get into trouble.'

SDC: This was in 1913 and already they were likely to regard him as being hostile?

JPG: Yes, they weren't very friendly.

SDC: That's an interesting point.

JPG: So I went on by myself, and the prince went down to Rome. I finally received the money in Vienna, and I wish I'd kept the envelope with its dozen forwarding addresses. It had been following me around.

SDC: All around Russia?

JPG: Yes. For the first time in weeks, I had money of my own. It so happened that on that very day, within an hour after I got the money, I got a cable from the prince stating that he'd lost money gambling, and if I'd received my money, could I please send him what I owed him? I was very conscious of the favour he'd done for me, so I sent him the money right away, and I think it left me with about $10. I was broke again until I got the next remittance from my father.

SDC: That was splendid of you, I must say.

JPG: I was about two hours with my pockets full of money. I spent about a month in Vienna, and it was wonderful. Vienna was under the Hapsburgs in the days of the emperor – very smart, very gay, and you had gold coins – you didn't need passports except in Russia and Turkey. It was the old world, not a regimented world. It was a world where there were practically no income taxes, national debts were very small, and people had never heard of gift taxes and capital gains taxes. People travelled and never thought of having passports with them. From Vienna I went to Izmir in Asia Minor and on to Athens, Greece. I was interested in Greek art and sculpture when I was there. I coveted some of the pieces I saw in museums, but of course, I had no money. I left the day before my birthday in 1913 [15 December] for Alexandria in Egypt. I was on a Romanian boat of about 3000 tons, and we had a very big storm in the Mediterranean. They had a big clock, and I would watch this clock and wonder if the boat would go another five minutes before it capsized.

SDC: It didn't capsize though?

JPG: Fortunately, it didn't. I probably wouldn't be here if it had. So, I spent the rest of December in Egypt. I finally left Cairo about the middle of February and got on a Cunard line boat to Gibraltar. I toured Spain, went to the Alhambra and to Seville. Then, it was on to Paris . . . and the beginning of the First World War.

SDC: You went back to Russia in 1935, didn't you?

JPG: Yes, to Leningrad and to Moscow.

SDC: Why did you go there particularly?

JPG: I was interested. I'm pretty much of a Cook's tourist, and I wanted to see Russia again. It had been twenty years, twenty-two years since I'd been in Leningrad and Moscow, and there had been a Revolution.

SDC: What sort of contrasts did you notice?

JPG: I was favourably impressed. The people were better dressed in 1935 than they had been earlier. I mean the poor people were better dressed. I didn't see anyone without shoes, and I had seen quite a few people without shoes in 1913. There was the impression of more well-being.

SDC: What did you do in Russia then in 1935?

JPG: I went to the Hermitage. It impressed me tremendously. It's the biggest collection in the world. They had 15,000 pictures – twice the pictures there are in the Louvre.

I went to the Trechikov Museum, too. I didn't go to Arcangelskoye, but I went to a palace, Cheremetiev's at Ostankino. The Soviets had taken good care of it. They have a long row of state rooms there with wax figures of lackeys, two to each door. They have the old ceremony which was adopted from France, typified by the Duchess de Berry. The Duc de Berry asked the Duke of Orleans to ask the king to give him equal precedence with his wife because he was very annoyed at the Tuileries Palace receptions. He would approach these double doors, which were shut, with his wife and the lackeys would open both doors. His wife would go on through, and then the lackey would shut half of one door, and he'd go through the other half-door. The reason was that his wife was the daughter of a king (the King of Naples), and he was only the son of the brother of a king, so he didn't get precedence.

SDC: Did you hear that amusing story on television the other night? Everybody turns up at the gates of St Peter, and St Peter says: 'All those who have been dominated by their wives, stand in one line. All those who have not been dominated by their wives, stand in another line. The line with all those who had been dominated by their wives stretched to the horizon. There was only one little man in the other line, you see. So, St Peter came out and said: 'What are you doing in this line?' And he replied: 'My wife told me to stand here.'

JPG (*laughing*): That's very good.

SDC: Anyway, Paul, what was the significance of this particular palace, the Ostankino?

JPG: Well, they use it partly for propaganda purposes because Cheremetiev was the richest man in Russia. At the start of the tour, they had some statistics showing that Count Cheremetiev owned land that was about half the size of Belgium. He had something like sixteen palaces in Russia, and as I remember the statistics, he had 27,000 domestic personnel in his various palaces. They showed excerpts from the accounts, for instance: 'Gave a dinner for Mimi – ten thousand louis' and 'Bad luck at baccarat – eight thousand louis.' In his library he had 30,000 books. Of those 30,000, there were only three on agriculture. Then, they have a ballroom there. You can see a peasant's hut right on the floor of the ballroom. They say: 'In the days of Count Cheremetiev, this was the home of twelve people.'

SDC: The Soviet contrast?

JPG: The contrast, yes. And then, for example, he had two hundred in the Corps de Ballet, which lasted for two hundred years.

SDC: When you visited in 1935, you were a well-known American businessman – known to be a capitalist. Did you have any difficulty getting a visa to go and visit?

JPG: Not a bit.

SDC: And when you got there, how were you received?

JPG: Very politely.

SDC: Did they show you around a number of their factories?

JPG: They gave me admission to the Red Army Club, which is a very luxurious clubhouse; I think it is the most luxurious clubhouse I've ever been in. The finest building and the finest furnishings.

SDC: Did they supervise your travels very much?

JPG: Apparently not.

SDC: You didn't feel that you were being followed or anything?

JPG: No, no.

# 8
# GETTY ON GETTY

Paul came through to the drawing-room from his study by way of a small ante-room where he kept his colour television set. He apologized for being a little late.

SDC: No, Paul, you haven't at all kept me waiting. I've only just arrived.

JPG (*turning towards the machine I had placed on the little French table beside his chair*): Is this recording?

SDC: Yes. We can switch it off if you want to.

JPG: No.

SDC: I wonder if you could point to any particular people you have known who've had a great influence on your character?

JPG: I think my father had a great influence on my life. I think his personality and his way of living, his 'lifestyle' as you might call it now, had an influence. Although I'm very different from my father, I've tried to remember his way of doing business. If I say so myself, he was a very honest man, and he made his money the hard way – risking it in the oil fields. I tried to bring that out in the introduction I wrote for *The History of the Oil Business* [by G. F. and J. P. Getty, privately printed.]

SDC: So you feel that you've held him up to yourself as an example of integrity. For, as Shakespeare said, 'noble minds keep ever with their likes'. Do you find that you have a great many interesting friends? You seem to have a great gift of friendship, really.

JPG: Well, I cherish my friends.

SDC: Who would you say are your principal friends, still alive?

JPG: The last of the friends of my own generation are thinning out fast. I had lunch as the guest of Arthur Rubenstein about three weeks ago. He is in London for the publication of his autobiography. It's a very good book about his young years. He's writing two books, one will be his young years, the other his later years. He told me that the first book was much easier than the second because practically everybody's dead at the end of it. He's eighty-six now, so nearly everybody he's writing about was as old as he was or anywhere from ten to fifteen years older. It could well be that practically nobody's alive that he was writing about!

SDC: So you think he had a freer pen about the earlier period?

JPG: Yes. It's an amazing thing, you know. I'm no lawyer, but I think that libelling a living person is very different from a dead person, isn't it?

SDC: Yes, indeed. The dead person hasn't got much chance of filing a suit against you.

> **On business**
>
> In the oil business there was a friend of mine who seemed to have all the essentials of success, all the characteristics that lead a man to business success: he was intelligent, hard-working, confident, ambitious, tactful; in fact he had got every characteristic, and I always thought he would be a great success. Actually he turned out to be a great failure.

JPG: No. I found that out when I wrote my *History of the Oil Business*. There was a friend of mine. He seemed to have all the essentials of success, all the characteristics that lead a man to business success: he was intelligent, hard-working, confident, ambitious, tactful. Actually, he turned out to be a great failure, and you can't write about a great failure. When he's still alive and well, you can't describe him as a failure. He might sue you; besides, it wouldn't be kind. So how do you express it? He turned out to be a minus quantity in the business. How do you express it so that the law is impassive?

SDC: What about letters? Do you reveal yourself a lot in your letters?

JPG: I've never been a letter-writer. I generally write very short concise notes. I send cables and have telephone conversations. I keep a journal which, of course, is generally very brief. It just mentions who I saw yesterday; occasionally a few remarks.

SDC: Have you kept them up to date?

JPG: Yes.

SDC: It might be of use for this book to look at that sometime. Presumably, of course, when you were getting married to your wives you wrote letters to them and other friends?

JPG: But I don't have the letters.

SDC: You feel that you have no letters that would throw a great deal of light on your character and activities?

JPG: No, I don't think so. You do have some of my early diaries.

SDC: Yes, and I've been picking up some interesting signs of your beginning to amass a fortune. I found that at the age of 11, you're beginning to count the fact that you've got 250 marbles and that within a month or three weeks, these have increased to 600 marbles. You refer to them as 'crockies' and 'goldies'. Can you explain that a bit?

JPG: They were different types of marbles. One kind would be worth three of the other kind. I think the 'goldie' was worth the most.

SDC: I noticed also that you were trying to build up a bank account and intended to open it when you had $12. In fact, you did start working very early to get some extra pocket money by going out and selling the *Saturday Evening Post*. I find it particularly intriguing that it was on your birthday, 15 December 1904, after receiving various presents, that you went out and spent that afternoon selling a magazine. I mean, not everybody works on his birthday. I remember working on my twenty-first on *The Times* newspaper from about five to midnight; and I resented it afterwards. But you obviously started working on your birthday.

JPG: Yes, I can still remember it. I was always very fond of the magazine. I read it myself. I think I took about twenty copies and delivered them around. I remember one man in particular who was very kind to me. I think he took the

*Saturday Evening Post* more to help me than because he like to peruse it himself. He'd buy a couple of copies – one for himself and one to give away. I think some others bought because it was a boy selling them who evidently was very keen to make a sale. It was very important to make a sale. I remember that the sale of the *Saturday Evening Post* was a matter of great importance. I hated to come back with unsold copies. I don't suppose I was ever more interested in any business than I was in my *Saturday Evening Post*.

SDC: I find it interesting. It does show a good deal of effort and enterprise at a very early age, because you weren't exactly dependent on that, were you?

JPG: No, I wasn't.

SDC: I mean, your parents were rich and you were doing this in order to start building up your own money?

JPG: Yes.

SDC: How long did you keep this up?

JPG: I think about six months. Also, there was something called the *Youth's Companion*. It was a weekly magazine, and I used to read it every week with great interest. I wanted some bound copies of it. The price was about $12 or $15, and my father gave me money for luncheon, maybe 25 or 30 cents. Those were the prices in the early twentieth century. I was very hungry, but I went without lunch and saved the money to buy the *Youth's Companion*. I still have those bound volumes of the *Youth's Companion* today.

SDC: Did you sell the *Saturday Evening Post* to make money to buy *Youth's Companion*?

JPG: Yes, that was part of the reason.

SDC: So you weren't able to bank a lot of your earnings from those sales?

JPG: No. I was very fond of a book called *Bears of Blue River*, which originally appeared in the *Youth's Companion*, and I have a copy of that book today.

SDC: Did you do any other work of that type?

JPG: I had a lemonade stand once or twice. I think I made a few nickels.

SDC: I notice you were reading a lot of Napoleon Bonaparte at the age of eleven.

JPG: Yes, I was. I think I read *Abbott's Lives*. He wrote the lives of about fifty famous historical figures.

SDC: Did you feel a great admiration for Napoleon at age eleven?

JPG: Yes, I did. I wondered about why he seemed to have a necessity of having so many wars, whether he couldn't have been more diplomatic and conciliatory.

SDC: Most people don't really start to apply their minds to why great conquerors have come unstuck until a bit later in life. It does suggest that you were very interested in the working of power at an early age, and this is rather prophetic.

JPG: Yes. Incidentally, I enjoyed the memoirs of Napoleon, the book you gave me of the Waterloo campaign.

SDC: Good.

JPG: Very well written.

SDC: Actually, it was written by Napoleon, but I translated it.

JPG: Yes, but your introduction was very good. Who is this man, the Frenchman, who wrote the other introduction to it?

SDC: La Chouque.

JPG: Was he a contemporary of Napoleon?

SDC: No. Would you like to have the first volume, too? I could get . . .

JPG: Thank you very much. You're a very generous person.

SDC: I understand you took a great interest in boxing at one time.

JPG: Yes. I always will remember taking boxing lessons with Mr Edwards, a very nice young man.

SDC: You very nearly became a professional boxer, I gather?

JPG: Yes. I boxed with Jack Dempsey.

SDC: You fought with Jack Dempsey! How did that arise?

JPG: Jack liked to box with me because he told me that he had more trouble with men who were smaller than he was. If a man was bigger than Jack, he was probably slower than Jack; and Jack liked to box or slug with a man who was slower because he'd have the advantage, whereas, if the man was smaller and lighter, he might be just as fast as Jack, maybe even faster. Well, I was pretty fast keeping out of the way of Jack's punches! (*Laughing.*) Once you've sampled one or two of them, you got right away!

> **On boxing**
>
> I boxed with Jack Dempsey. Jack liked to box with me because he had more trouble with men who were smaller than he was. If a man was bigger than Jack, he was probably slower than Jack; and Jack liked to box or slug with a man who was slower because he'd have the advantage, whereas if a man was smaller and lighter, he might be just as fast as Jack, maybe even faster. Well, I was pretty fast keeping out of the way of Jack's punches! ... I saw him take part in some of his world title fights. I think he was the best. The shape he was in during his early fights – he was unbeatable. I don't think anybody could ever have beaten him.

SDC: I bet, yes. How much training had you had by that time?

JPG: I was in my late twenties.

SDC: Had you seriously thought of taking up boxing?

JPG: I had been doing boxing occasionally. I kept up the practice regularly.

SDC: How did you come to know Jack Dempsey?

JPG: I met Jack at Crystal Pier about 1918 or 1919, and I liked him very much. He was a splendid fellow and still is.

SDC: Did you often box with him, give him practice?

JPG: Yes, and I saw him take part in some of his world title fights. I think he was the best. The shape Jack was in during his early fights – he was unbeatable. I don't think anybody could ever have beaten him.

SDC: Did he pull any punches when he was practising with you?

JPG: No, he didn't. (*Laughing, amused at the recollection.*)

SDC: You seem to have survived. You don't appear to have cauliflower ears or anything as a result of it. Nor does your brain seem to have got addled by being punched.

JPG: No, I was moving very fast. When Jack was after you, you had to move very fast.

SDC: Did you box anybody else apart from him?

JPG: Yes. I boxed Stanley Ketchell one time. It's said of Stanley that if he hit a man and the man didn't fall down, Stanley would walk around back of him and see what was holding him up.

SDC (*laughing loudly*): He must have been very self-confident.

JPG: Stanley was an interesting man. I understand he was brought up in Grand Rapids, Michigan. When he was nineteen he had beaten every man in Grand Rapids who thought he could fight. Not that Ketchell was quarrelsome, but if he heard that a man was handy with his fists, he'd approach him and say: 'I understand you like boxing; would you like to box with me?' If the man agreed, why then, Ketchell would take him on. If the man said: 'No', then Ketchell would thank him and walk away. But for those who wanted to sample Ketchell's boxing, he established supremacy. Then, he went up to Montana and took a job as a bouncer in a dance hall. The first night he was on duty, he threw out forty men, one after the other. This is what he told me. (*We were both laughing a lot by now.*)

> **On boxing**
> I boxed Stanley Ketchall one time [the world middleweight champion]. It's said of Stanley that if he hit a man and the man didn't fall down, Stanley would walk around back of him and see what was holding him up.

SDC: Did they get any more customers for a fortnight?

JPG: The place was very quiet after the first night. He went on to become middleweight champion of the world.

SDC: I see. I didn't realize that. You came up against some pretty tough opposition. What made you give it up?

JPG: Well, I thought there was more promise for me in the oil business.

SDC: There's something interesting here. Obviously, you

had a fairly pugnacious spirit. You don't go in for boxing world champions unless you're fairly aggressive in action.

JPG: I was never interested in picking fights, and I always avoided picking fights. So has Dempsey. I've been around Dempsey when youngsters have challenged him, but, of course, Jack had nothing to gain by fighting with them. If he fought with them and knocked them out, the papers would say what a terrible thing it was for a professional boxer to knock out a poor college student who didn't know anything about it. I think a lot of those fellows knew that Jack wouldn't accept a challenge. They just wanted to make a point by saying: 'Well, Jack, how would you like to have a round with me?' Jack would laugh and say: 'No, thank you.' But he was never quarrelsome.

SDC: I didn't mean that you were quarrelsome, but boxing against people of that calibre showed a certain ability to hold your own in the ring against very tough opposition. Presumably, this came out in your business activities later on in a different form.

JPG: Maybe, yes.

SDC: Referring back to your early diary, you tell of a whipping your father gave you for calling him a 'doggoned fool' and saying that the chambermaid can go and soak her head in Jip's mouth [Jip was the dog]. Can you remember the incident?

JPG: I think my father had told me not to bring the dog into my room, to leave him outside the hotel. Obviously, I didn't want to leave him outside, so I brought him into my room unknown to my father. He wouldn't have known except the chambermaid came in, discovered the dog, and complained to my father.

SDC: And your father gave you a hiding. Did he do this often?

JPG: Very seldom.

SDC: So he must have felt particularly angry about the dog?

JPG: Yes.

SDC: On the whole, you seem to have been a very lively young man at age eleven. Once you rushed about shouting:

'Fire! Fire!' at midnight. Was this to get everybody out of bed in a panic?

JPG (*laughing heartily at the recollection*): Yes. Their language when they found out it was a hoax! I thought they needed livening up!

SDC: I see. Let's hope the house didn't burn down later on because you were shouting 'Fire!', and they didn't believe you. It never happened again?

JPG: No, but it was a good fire drill for them.

SDC: Is this practice one you have followed in your business activities sometimes – giving people false alarms to keep them on the move?'

JPG: I've given them a dry run occasionally.

SDC: Yes, it works, doesn't it? What more do you remember of this early time? Your diary breaks off after you were eleven, and we don't get much more until you were much older and travelling around Europe and so on.

JPG: I was very attached to my dog, Jip. It was a typical relationship, I suppose, that I considered Jip a great treasure. I couldn't have put a value on Jip.

SDC: What kind of dog was he?

JPG: He was a mongrel. Small, about that size (*making a gesture with his hand of about 18 inches – 45 centimetres – above the floor*). I always remember he was always hungry, and I made arrangements one time with the cook at the hotel to give him leftover steaks. That was the only time he had more than he could eat. I came out with this tremendous dish of meat. He ate until he was swollen to about double the size!

SDC: That was a nice thing to have done for him.

JPG: I remember another time. I liked to walk up the railway track about 3 or 4 miles [5 or 6 kilometres] with Jip following me. There was a bridge across the Caney River. I had no difficulty in stepping across the ties, and I wanted Jip to follow me; but somehow Jip missed the foothold and fell into the river. I though he might get carried away, but he managed to swim ashore, and he joined me. He wouldn't cross the bridge anymore, though; I had to carry him across.

SDC: Any other episodes connected with Jip?

JPG: Yes. They used to drive herds of cattle through the streets of Bartlesville. Surreptitiously, I set Jip on one of the half-dozen cowboys shepherding the cattle. Jip scattered the cattle, cowboys, and dogs. There were some rare oaths.

SDC: I notice you've always been very fond of dogs. You referred to Jip a lot in your diaries. You had an Alsatian you were very devoted to.

JPG: Yes, Shaun. I was devoted to him, and he was devoted to me.

SDC: Very sad when he died. Since then, have you had any special dogs?

JPG: No, I haven't a central dog. I haven't ever found a replacement.

SDC: I remember lunching here once with Juliet [my wife], and there was a very charming woman, Baroness von Alvensleben. After lunch, we went out to see a lion of yours. I remember getting into the cage with that thing. We must have had a rather good lunch, I think! The lion got its teeth around my trouser leg and you said: 'I think I owe you a new suit, Somerset.' (*Getty laughed.*) Fortunately, he was well fed so he didn't really penetrate the trousers or anything else. Have you still got the lion?

JPG: I still have. I've got another one now; I've got two.

SDC: They're really not here as guard lions, are they, so much as pets?

> **On relaxation**
> I think that I have a great gift for idleness and for beachcombing. I think I could have been a good beachcomber.

JPG: They are pets.

SDC: Do you go out and watch them being fed and that sort of thing?

JPG: Yes, it's a diversion.

SDC: What other diversions do you have? Do you have a lot of hobbies?

JPG: I used to be interested in weight-lifting, and I've

always been a great beach man. I like roving at the beach, swimming, sun-tanning, surfboard-riding.

SDC: So you can, in fact, relax quite well?

JPG: Yes. I think I have a great gift for idleness and for beachcombing.

SDC: Were you ever musical?

JPG: I used to play the piano, but not well. I can remember 'Dolls Dream'. In fact, I used to play it often and could probably play a few bars of it now.

SDC: Were you ever interested in the language arts?

JPG: Here I have a story. I was a guest of King ibn Saud at his palace in Riyadh. The palace is vast, and I got a reputation of being very fluent in Arabic, which I never was. An Arabist, an American who spoke very good Arabic, was at the bar in the house and wanted to get a pineapple drink. He couldn't think of the word for pineapple in Arabic, so I said 'ananas', which happened to be the word. He gave me one long look!

SDC: He thought you were a linguist. What languages do you speak?

JPG: German, French, Spanish. I speak some Italian. I'm not proud of my Italian. I have great admiration for people who speak fluent Spanish and then switch to fluent Italian.

SDC: I find it almost impossible to remember Italian after I've been speaking Spanish.

JPG: I say 'gelo' instead of 'giaccio' and say 'como' instead of 'come'.

SDC: Have you found that it's been useful speaking these different languages?

JPG: I think so. You lose a great deal in a foreign country if you don't speak the language. Before the war I was in Hungary, in Budapest, and I didn't speak any Hungarian. I felt like a fish out of water. If you're in Paris and you speak French, you feel as though you've got the freedom of the city, so to speak. I was in Constantinople before the Revolution, and I don't speak any Turkish except a couple of words. I think there's a lot of difference.

SDC: How do you feel about getting older? Are there any advantages?

JPG: Well, growing old is something like Robert Browning's poem, 'Along With Me':

> Grow old along with me!
> The best is yet to be,
> The last of life, for which the first was made:
> Our times are in His hand
> Who saith, 'A whole I planned,
> Youth shows but half; trust God: see all, nor be afraid!'
> (Robert Browning, *Rabbi ben Ezra*, i)

SDC: Splendid.

JPG: I don't know whether growing old is good or not. I tell you, it makes a difference when you get something the matter with you. You're twenty years younger than I am, and you've got twenty years more strength. You get ill, and you find your strength is cut into, but you have more strength to start with. When I get ill, I haven't got as much strength to cut into. The result is, with the half that's left to you, you're just about able to get up the stairs!

SDC: As they say, the veil is wearing thin. But I don't altogether agree with you. I may be twenty years younger than you, but I've been astonished being associated with you in this way. You have an enormous capacity for work. So although you have a theoretical lead over me, you don't seem to be showing it!

I told him that it would be possible for him to see a draft of our conversations soon. He seemed rather relieved about that since he was not getting any younger. He was over eighty and feeling rather frail this day. He had some difficulty getting out of his armchair when he got up to see me off. I had to help him. He said it was a touch of arthritis.

# 9
# WAR

This conversation took place on 17 October 1973 in the drawing-room at Sutton Place. Paul mentioned the date as the anniversary of his father's birth in 1855.

SDC: Shall we talk about the Middle East and the latest confrontation? [In 1973, war broke out between Egypt and Israel when Egypt launched a surprise attack on Israel during Yom Kippur.]

JPG: If you'd like.

SDC: Do you feel there is likely to be a world war rising out of the present situation?

JPG: There could be. The principal danger of a war would be if the Israelis get beaten decisively and pushed back towards the sea. I think there'd be some sentiment on the part of some people to freeze the line, the divisional pre-1967 line, but the Arabs didn't agree to that. It might be an innovation.

SDC: There is some evidence that we're now facing a *jihad*, a sort of holy war. The Arabs are in one of their rare spasms of religious fervour – this all started during the month of Ramadan. You've been out in the Middle East during the month of Ramadan; you know how tempers rise. Passions are excited when people don't eat or drink from dawn 'til dusk. Also, it seems that for the first time in many years, all the Arab nations seem to be uniting in this particular fervour to attack Israel, don't they?

JPG: Yes, they do.

SDC: The Israelis feel equally indignant because they were attacked on their Day of Atonement. I would say there are all the elements here for a very widespread clash in the Middle East. You've heard on the news today that there is talk of

Russian tanks and equipment pouring in. America, too, is said to be putting some missiles aboard an Israeli plane. All this could lead to widespread escalation.

JPG: I remember when I arrived in Paris about the end of May 1914. I went to the Grand Prix and was sitting in a private box near the President of France. During the races I noticed considerable excitement around the President. A messenger had arrived and had handed a dispatch to him. I didn't know what was in the message but I could see that it was important. It had been news of the assassination of the Grand Duke and Duchess in Sarajevo, which was the start of World War I.

SDC: How interesting. What happened immediately on receipt of that news in Paris? Did it penetrate that this was a decisive event?

JPG: People were worried about it, but nobody thought there'd be war. There was no more a feeling that there would be war than we see now, that there would be war between the United States and Russia.

> **On the communist position**
>
> I think that it is a fundamental weakness of the communist position that there is Russia on the one side and China on the other. I think Russia would be very opposed to getting into a war with the United States. China would possibly join in because China has great territorial claims against Russia. I would think that war between the United States and Russia would not result in, at least at the start, nuclear exchange. I think it would be restricted to conventional weapons. China's getting stronger every day, every week, every month. China will probably pin down a lot of Russia's forces on the Chinese frontier. Until the communists sort out their differences and there's a reconciliation between Russia and China, it's like a man with two legs that won't synchronize – he can't go backwards and he can't go forwards.

SDC: At the rate we're going, we're going to be very short of fuel for winter heating and other purposes. What do you see of the future yourself?

JPG: I see a fundamental weakness of the communist position – there's Russia on the one side and China on the other.

I think Russia would be very opposed to getting into a war with the United States. China would possibly join in because China has great territorial claims against Russia. I would think that war between the United States and Russia would not result in, at least at the start, nuclear exchange. I think it would be restricted to conventional weapons. China's getting stronger every day, every week, every month. China will probably pin down a lot of Russia's forces on the Chinese frontier. Until the communists sort out their differences and there's a reconciliation between Russia and China, it's like a man with two legs that won't synchronize – he can't go backwards and he can't go forwards.

SDC: This is assuming, of course, that the war doesn't break out between Russia and America due to the kind of confrontation we recently had. Do you think the Russians were influenced by this threat from China in their rear?

JPG: I think they were heavily influenced by China.

SDC: Do you think they will eventually come into the war themselves? The Russians and the Chinese?

JPG: It's possible.

SDC: Russia is such an enigma now. They've been preparing for war on a very big scale, haven't they? When you get a showdown like this, you suddenly find them in a very strong position.

You're a bit pessimistic about it, aren't you?

JPG: I'm not exactly pessimistic. I think the odds are five to one that there won't be any spread of the war to major powers, but even odds like that make you feel uncomfortable. I mean, it's not a hundred to one shot; it's not a million to one shot. If I were setting the odds, I'd set them at one chance in five.

SDC: Well, that's pretty bad.

JPG: If somebody pointed a gun at you and told you that the statistics were that there was only one chance in five that it was loaded and pressed the trigger, I don't think you'd take very much comfort out of the statistics.

SDC: No, there was a cartoon the other day of six fellows in a row and a revolver with six possible bullets in the chamber. The first fellow took a shot, and he was all right. He passed it on to the next, held it to his temple, and fired. He was all right. And it went on. Five of them got past. There was only one man on the end of the line. He turned round and shot the other five all through.

JPG (*laughing loudly*): It's one way to get out of it.

SDC: By the way, did you enjoy the actual life of the Middle East apart from your business activities? You must have spent quite a lot of time out there, and you were presumably sitting cross-legged in tents with the sheikhs and having meals of roast lamb piled up with saffron rice, eating with your hands.

JPG: Yes, it was fascinating. I enjoyed it. I liked the Arabs. So splendid business-way, too.

SDC: Fine. You were very acceptable to them because of not having any Jewish blood.

JPG: Yes.

SDC: You didn't come up against any of their prejudices against the Israeli element? Did you have an advantage in your dealings with the Arabs, do you think, because you are not a Jew.

JPG: Well, it didn't hurt me any. However, I think that the Arabs make a distinction between a Jew and a Zionist; in other words, they don't necessarily have any hatred of a man who is a Jew because they realize that you don't have much choice about your race. But Zionism, that's different. So if a Jew is not a Zionist, I understand that he's well accepted.

SDC: Any more thoughts?

JPG: I've seen in my lifetime the passing of the generations. I can remember particularly the GAR, Grand Army of the Republic, conventions. They used to have an annual convention, and they'd march through the streets, bands playing, thousands of men marching through the streets, veterans of the Civil War. And where are those men marching today?

What time will you be here tomorrow?

SDC: Three o'clock, it that suits you.

JPG: That'll be all right.

We got up and moved towards the front door. Getty indicated

a pile of heavy, leather-bound Golden Cockerel Press volumes of my Napoleon Memoirs which I had brought to show him.

'You're heavily laden. They're very handsome. Beautifully bound.'

There was a great clanking of bars.

'You've got only one more bar, I think,' I said. Getty stooped to unbolt the last one. We shook hands.

'Thank you and goodbye, Somerset. See you tomorrow.'

'Goodbye, old chap.'

# 10
# INSIGHTS

### *Getty on . . . Attorneys*

I wanted to get a good attorney, so I went to Charlie Mitchell, who then headed the National City Bank. I asked him who a good attorney was, and he named a prominent New York man. I went to see him and was impressed by him, but I noticed that every time I asked him a difficult question, he pressed a button and a young Jew came into the office. This lawyer would ask the young man a question and the young man would reply without looking it up at all. He'd give him what the law was and the chances that the judge would understand the law and then the chances that the judge would apply the law.

In each case you apparently need three things. In the first place, you have to understand what the law really is, and then you have to estimate whether the judge is competent enough to understand what the law really is, and then you have to estimate whether he will apply the law.

It's not just a question of whether you've got the law on your side. Does the judge know that the law is on your side? This lawyer, Dave Hecht, was a mental athlete. Like a friend of mine once said to me: 'Paul, I don't fear you, but I fear your attorney!'

### *Getty on . . . The environment and pollution*

*To what extent did the oil prospectors obliterate cities?*
The Athens area – so-called – was a state development cut up into town lots and streets with curbs, but building hadn't started, with the exception of maybe every twentieth lot. So, it wasn't a situation where drilling oil required houses to be

demolished. A house in Athens at that time might have had a rental value of about $40 a month, maybe $50; whereas, a really good oil well might have the value of $2000–3000 a day. So, they didn't waste much time in choosing what they wanted. An individual couldn't have kept a view anyway because he couldn't see anything but oil derricks.

I did have one problem there. We built part of a derrick on the sidewalk, and it seems I was notified by the city that I'd have to move it. Well, actually, it wasn't much of a sidewalk. It wasn't paved. It was just dedicated as a sidewalk... but I moved it [the derrick]. It violated regulations.

## *Getty on . . . The oil shortage*
*(Taped in 1973)*

I was asked by Columbia Broadcasting if the oil crisis is real, and I said it was, and whether it was expected, and I said I'd been expecting it for fifty years. How are we going to deal with it? We're going to have to try to increase production, and of course, the people in our government have done everything to bring it about.

If you'd been writing a book on how to bring about an oil shortage, you'd have a very good imagination, know the subject thoroughly, and you'd have twenty different ways to bring a shortage about. We've done every one of them. I mean it's just like if you wanted to get rid of all your friends, and you've got a good vocabulary. If you single out each particular friend for vocal vitriol, you'd probably get rid of a few friends. That's what we've done.

In the first place, we've made nuclear power as short as possible. We've had the environmentalists and the government commission that regulates these things: people who are very hostile to nuclear plants. Nobody wants a nuclear plant next door, so it's almost impossible to get a permit to build a nuclear plant. They're changing the regulations frequently on the ones that are being built. We've done a very good job about stifling the nuclear industry.

Then, our automobiles use about 15 to 20 per cent more gasoline than they used to because of clean air regulations. We're using a tremendous amount of energy on air conditioning and heat. We've kept the price of natural gas artificially low for years, which discouraged people from going out and drilling wells to try to discover more gas. Then there are the Middle East conflicts. We have a very good record of bringing it about.

Fields such as those in Alaska and the North Sea will not be sufficient to catch up on the demand. The demand is increasing so rapidly that a single field, no matter how big, is not important any more. We'll have to make it possible to get nuclear plants commissioned, and we're going to have to take some of the controls off the prices of oil. The way to increase production is to increase the price; to have the price at about the international price, that's guaranteed to throttle production. The oil we're getting now is more expensive to produce. Ultimately, of course, we'll run out completely, and the high price won't have any effect. But, we haven't tried the high prices yet, and I think we'd get a temporary fillip by a higher price. It might carry us on for a few years. We won't have a surplus of oil, but we'll have more than we have now!

Are there any more fields of oil in the world to be found? Yes, there must be, particularly off-shore.

## Getty on . . . Politics

### DIPLOMATS

I've never been able to gather why it is, but the British are the same way as the Americans and other nationalities. They'll send a man to Tokyo where he learns some Japanese and gets to know the people. After he has been there for two or three years and is a better man as an ambassador with better judgement of the Japanese, he's suddenly transferred to Peru. He serves there three or four years, learning Spanish and the background of the Peruvians, and he's transferred to Greece. Then, he goes through the process again.

DICTATORS

Theoretically, I would admire a dictatorship providing that the dictator was an efficient and reasonable man and that he behaved himself. Unfortunately, many dictators haven't been that way. I'm often reminded of Winston Churchill's statement that democracy was the worst possible form of government until you compared it with anything else. Speaking of Churchill, his son-in-law asked him who he thought was the greatest statesman of all time. Churchill replied: 'Mussolini'. Rather surprised, the son-in-law said: 'Sir, why do you say that?' Churchill pontificated: 'Because he shot his son-in-law.'

WATERGATE

I think the idea of Nixon's impeachment is very vicious. The Democrats are nearly two to one in both the House and the Senate, which of course makes it dangerous for a minority president. Well, if there's any justice, he certainly could not be convicted of high crimes and misdemeanours. In the first place, I wouldn't say his so-called obstruction of justice was a high crime or misdemeanour.

> **On judges**
>
> Anybody is guilty of obstruction of justice. As a matter of fact, the worst people are the judges in obstructing justice, because supposing the judge says to you in America – which they did in these Watergate tapes – 'I'll give you four years or two years if you cooperate and tell me everything you know, but if you don't I'll give you forty years.' Well, now, that is obstructing justice, because that is taking away a man's civil rights. He doesn't have to testify against himself; that's the Fifth Amendment.

As a matter of fact, the worst people are the judges in obstructing justice because they take away a man's civil rights. They'll say, 'I'll give you four years or two years if you'll cooperate and tell everything you know; but if you don't, I'll give you forty years.' He doesn't have to testify against himself; that's the Fifth Amendment. There's a judge trying to bring him about where he'll testify against himself, against others. So, that's obstruction of justice just as much as anything else. Anything that interferes, that tips the balance one way or the other, is obstruction of justice, theoretically or practically.

Every lawyer is guilty of obstruction of justice. For instance, Max Steuer was a great man for it. I saw him in an income tax case when he was defending Mitchell, the president of the National City Bank in New York, against the government. They had an ironclad case against Mitchell, a tremendous amount of evidence, forty witnesses who swore he was guilty. Steuer capitalized on that, saying: 'Is this a Russian trial? Where there are 5000 people who swear so-and-so is guilty according to Stalin? Nobody could be that guilty!' No legitimate prosecutor could accumulate so much evidence; you see, it's a rigged trial. The jury went out and pronounced him innocent. Steuer certainly obstructed justice because he capitalized on the weaknesses of the jury.

> **On the Nixon tapes**
>
> If you're going to have a situation where the President of the United States can't have a confidential conversation with, let's say, Brezhnev, how is he going to function? You can't have a situation where the Prime Minister has a private conversation with the Queen and then a recording is made of that conversation, and then some court broadcasts the conversation between the Prime Minister and the Queen. I mean, the Prime Minister cannot very well function, and the Queen cannot function.

It seemed to me that Nixon had to get rid of the prosecutor, Cox, and the new Attorney General. I think he made a reasonable offer by offering to prepare a summary of the tapes. If you're going to have a situation where the President of the United States can't have a confidential conversation with, let's say, Brezhnev, how is he going to function? If the courts claim that these tapes be pulled out and the contents be aired and decided upon, he can't function very well.

HIS OWN POLITICAL INVOLVEMENT
I've never been drawn to politics myself. I always felt that others could probably do better than I could. I thought about going into the Foreign Service as a young man, but I think there was a prejudice. I was one of the rich people, and thirty or forty years ago, being a millionaire was a handicap. Now you see Rockefeller, Harriman, the Kennedys. I'd probably have a better chance now if I ran for president so far as the money situation goes. The theory in America forty years ago was that you had to be born in a log cabin.

## *Getty on . . . Smoking*

Paul Getty was an inveterate smoker and found himself late one night in a hotel room when it was raining hard. He started looking in his pockets for a cigarette. To his dismay, he found that he did not have one on his person or in his room. He knew that there was a tobacco stall three blocks down the street, and although he had undressed and was getting into bed in his pyjamas, he decided to dress again and go out into the night, cold and rainy as it was, to get some cigarettes.

While he was putting on his raincoat, he suddenly asked himself if he really ought to have his head examined. 'What am I doing,' he thought, 'at one in the morning, going out into the cold, wet night in order to get a cigarette?' He decided then and there to give it up.

(Paul Getty, *How to be Rich*, Playboy Press, Chicago, 1965)

This is the account Paul related in his own memoirs. What he told me is still more interesting. He did not just give up smoking then and there and never think of touching a cigarette again; he decided that this was beyond human capacity. Instead, he set himself the task next day of seeing if he could go for another day without a cigarette, and having survived that day, said: 'Well, I'll try and see if I can survive another day,' and so on. Finally, he broke himself of the habit altogether. To his dying day, he did not smoke at all. This strength of character manifested itself in his daily life.

## Getty on... Writing and writers

Paul Getty was a fluent writer who published at least seven books. As an author, considering that he was by no means dependent on his writing for money, his literary output alone would command attention. He often said he would have been a writer if it had not been for his business preoccupations. Certainly, he had 'the root of the matter in him,' as Rudyard Kipling expressed it. He did not cultivate it and let it grow as he should have.

He collaborated with Ethel leVane on a book called *Collector's Choice*, (W. H. Allen, 1955), providing the later section called 'A Journey from Corinth'. In that he recreated the life of a young couple who take the statue of Hercules to Italy. It is in fact a short novel and shows his considerable knowledge of classical times.

Books he wrote unaided include *The Joys of Collecting* (Country Life, 1966), which reveals his immense knowledge of the background of his works of art. He was also the painstaking author of the massive tome *The History of the Oil Business of George F. Getty and J. Paul Getty*, privately printed. When I borrowed it for reference, he said: 'Don't lose it. I could not get $5 for it, but I would not part with it for a thousand.'

Another privately printed book is *Europe in the Eighteenth Century*. Here we find an extremely vivid and readable account of the life and art of the period, and it deserves wider circulation. Once again, it reveals his immense erudition and discriminating taste. He writes in it:

A piece of furniture can be a great work of art and may command a high price. Some people whose culture is not very deep are not aware that furniture can be of great artistic importance. The throngs of visitors to the Wallace Collection in London gaze reverently at the pictures in some of the rooms and hardly glance at the furniture. Yet, the furniture is worth more than the paintings in some of the rooms. A collector needs a large purse. His beautiful paintings should be matched by furniture and carpets of equivalent quality.

During our conversations at Sutton, Getty said of the book:

I wrote this book in longhand. It was a terribly difficult subject because it's surprising how little we know – or at least I knew – about the eighteenth century. For instance, how did they clean their teeth? They didn't have toothbrushes; they used to swab their teeth with cloths! And you speak about prices . . . you say the best French furniture was high-priced. Well, how high-priced was it? What were a pair of corner cupboards, let's say those by DuBois, worth in 1780, for example? I found I had no idea. What time did they eat lunch . . . or dinner? They normally had dinner about three in the afternoon. I made up my mind when I was halfway through this book that if I ever wrote another story, I'd write a novel about the oil business – the contemporary oil business. There, I would know the answers to all the questions – wouldn't have to do all the research!

# 11
# PERSONAL PAPERS

### *Getty's Unpublished Diaries*

J. Paul Getty kept diaries most of his life, beginning in 1904 at the age of eleven. His entries at the various stages of his experiences shed a great deal of light on the shaping of his character and his future attitudes and opinions. What were the important events in Getty's mind that he wanted to remember? Here are but a few excerpts from diaries kept when he was a youth and during his mid-forties.

### 1904

During this year we begin to see the emergence of an instinct for acquiring property and a more mischievous side as seen through his boyish pranks.

*17 February*: Fine day. I played in the morning and afternoon. In the evening Papa gave me a whipping for saying he was a doggoned fool and the chambermaid had better go soak her head in Jip's mouth. (*Jip was Paul's dog and good friend.*)

*22 February*: Fine day. In the morning I went over to the barn and got the guys mad by shouting at them and setting Jip on them. All the boys were playing marbles here and some of the girls.

*28 February*: Fine day. In the morning I rode. In the evening I played Scout, creeping up to a person in the dark without them seeing me. Jip came along and barked at the moon and the bonfire.

*10 March*: Fine day. In the morning I started to school for the first time in six weeks. I'm in Miss Pelton's room. In the afternoon Harry, William, and I played marbles for keeps. I skinned a flint.

*15 March*: Fine day. Went to school in the morning and afternoon. In the evening Harry came up, and we rolled for marbles with some of mine. I now have about 275 marbles.

*17 March*: Sloppy day. In the afternoon I said I would snowball a girl. She went and told James, and he chased me in the house and tried to get in. I hid under the bed.

*23 March*: Fine day. In the morning I went to school. In the afternoon I went to school and after school I played marbles with Harry. I skinned about twenty crockies and two goldies.

*27 March*: Fine day. In the morning I read *Napoleon Bonaparte* and went to Sunday School. In the afternoon I played with Harry and Ruth. In the evening I read and counted my stamps. I have 300 and 200 traders.

*29 March*: Cloudy day. In the morning Harry and I played marbles. I skinned twenty crockies. In the afternoon we went out driving with Prince. In the evening I counted my stamp collection. I have 305.

*1 April*: Fine day. In the morning I played marbles. In the afternoon Harry and I went over to the library to get a book, but it was closed and we had to go back. I fooled people and rang doorbells.

*10 April*: Fine day. In the morning I went to Sunday School. In the afternoon I read *Napoleon Bonaparte* and took a walk. In the evening I popped corn. There isn't anything to do on a Sunday.

*11 April*: Fair today. In the morning I went to school. After school I went downtown with Harry to get a book. In the evening I read and counted my marbles. I have 500 crockies.

*20 June*: Fine day. In the morning I went up to Miss Pelton's for my lessons. In the afternoon I went down to Papa's office and earned a dollar. I'm going to put my money in the bank as soon as I get twelve dollars.

*23 June*: Rainy today. I went up to Miss Pelton's and got there just as it started to rain. In the afternoon, Papa gave me 50 cents for a bicycle tag. I got it and put a dollar and ten cents in the bank. I have 15 dollars, 58 cents now. (*So, we know the extent of the J. Paul Getty fortune as of 23 June 1904!*)

*26 November*: Fine day. In the morning I went downtown with Mama. I got a new suit. In the afternoon Mama and I went to the Metropolitan Theatre. The play was *The Billionaire*.

*8 December*: Fine day. In the morning I went to school. In the afternoon I played with some boys. In the evening I sold the *Saturday Evening Post* and read *Both Sides of the Border*.

*Thursday, 15 December*: Fine day for my birthday. I got a bank and a box of writing paper, also a comb. In the morning I went to school. In the afternoon, I sold the *Saturday Evening Post* and got fifty cents.

(*Here we have a boy of twelve who, on his birthday, goes out selling the* Saturday Evening Post, *no doubt to be added to the bank account which was to be started when he reached $12. Paul told me – page 99 – that the business enterprise into which he turned the greatest effort in his whole life was selling the* Saturday Evening Post!)

*24 December*: In the morning I went downtown with Mama. In the afternoon I went to Cooks for my boxing lessons. In the evening I read and hung up my socks (*to be filled by Santa*).

*25 December*: Finally, I got a box of candy, a book, a game, a tie, and a lot of fruit. Mama got a diamond ring. I went to Sunday School. I also went out driving. In the afternoon I read and wrote. Papa gave me a dollar. Whee!

## 1938

On these pages of Paul's diaries, we catch a glimpse of Getty as he was at forty-five and touring Europe. Here we see a more sophisticated fellow – a man who has travelled a long way from the oil rigs of Oklahoma.

*Friday, 2 September*: Awakened late. A heavenly day. Sun shining in a cloudless sky. I took a sunbath in my room and admired the view until lunchtime. The view from Chamonix Valley is almost unrivalled. The difference in altitude between my window, 1000 metres [3300 feet], and Mt. Blanc, whose summit is only six miles [10 kilometres] away, is 3,800 metres [12,500 feet]. The great glaciers and the unique needles, Dru, Grande Charmoz, Grepon, Blaitière, and Midi, are but two to four miles [3 to 6 kilometres] away. The needles are fantastic and unbelievably steep. I never have seen their equal. Everything is terrifyingly grand! The north side of the valley – Brevant, Plan Praz, and Flegière – would be considered colossal anywhere else. The Brevant ropeway is still working. Cars look like airplanes. Mont Blanc itself is all that a monarch should be. Perfect weather the entire day. Spent the afternoon walking around and admiring nature in her grandest mood. Evening cold. Had dinner at the hotel. Still the dozen guests. Drank too much Asti Spumante and, to crown it, ate a lot of chocolate. Read until late and then couldn't sleep because of indigestion. Finally went to sleep about four.

*Sunday, 4 September* (*travelling about Switzerland on his own*): My 1937 Lincoln Zephyr runs like a charm. Had the feeling of the old days during a short walk. Drove on to Berne over nineteen miles [30 kilometres] of good road. Berne is a larger edition of Fribourg. Its old-world atmosphere is enhanced by arcaded street. Telephoned Fini and found Ronny all right. Spent the night in Berne at the Schweitzerhof after a good dinner at the Bellevue Palace. (*Fini was his third, and at this time former, wife, Adolphine Helmle. Ronny was their son.*)

*Monday, 5 September*: Drove to Interlaken and up the grand and glorious Lauterbrunnen Valley to Lauterbrunnen, eight miles [13 kilometres], and then past the glorious Staurbach to the Trummelbach Falls. Took the elevator up. What a grand spectacle. It is a moving example of water's power of erosion. I estimate there have been ten thousand years of erosion since the glacier melted. Next, I drove on to my favorite Grindelwald and the Bear Hotel. Found the same pleasant atmosphere and twenty guests, mostly English. One pretty girl from London, Eve Dunne, was my dancing partner for the evening. Went to bed late after reading Baedeker. Telephoned Teddy, but she was at the opera. (*'Teddy' Lynch was his fifth and last wife. He married her in Italy a year after this entry.*)

Telephoned the Lotti at Paris and had cable re Pierre read (*reference to his purchase of the Hotel Pierre in New York*). Ordered mail forwarded to Schweitzerhof, Basle. It is hard to give the palm for grand scenery at Chamonix. Grindelwald and Zermatt are all glorious contenders. I must make a choice. It is Grindelwald, Chamonix next, then Zermatt.

*Wednesday, 7 September* (*the day he bought the famous Ardabil Carpet*): Up late. Lunch at the Schweitzerhof. Afterwards, I telephoned Lowengard in Paris and bought Duveen's carpet for fourteen thousand pounds, after offering thirteen thousand pounds and being refused. Drove along the shore of Lake Lucerne to Fluelen. The day is rather misty, and effect one of absolute enchantment. Fairy peaks. Back to Lucerne after finding the hotel at Axenstein closed. Spent the night at the Carlton. Not as good as the Schweitzerhof, but the same price for a room.

*Friday, 9 September*: Up at 11 a.m. Car's springs greased for two francs. Ronny was with me all the time. Said goodbye to the dear boy and to Fini and left at 2 p.m. for Basle. No scenery, so made plenty of speed on a fine, swift road. Picked up mail at Schweitzerhof, Basle. No inspection of luggage at frontier. Great courtesy. In Germany, I headed for Baden-Baden on a greatly improved road. People look happy. Had dinner at the Stephanie and then read my mail. Walked along the brook for three minutes to the Kursaal. Won sixteen marks at roulette on four straight wins. A beer at the bar is one mark. The gambling rooms are beautiful, French style, baccarat lustres. No foreigners to be seen. Roulette only this evening at two tables. White chips at two marks. Betting limit is from two marks to a thousand marks. Mostly two mark and five mark bets made. A few red chips, twenty marks, were bet. Home to bed. I have a wonderful suite with hallway, sitting room, bedroom, bath and veranda for twenty marks.

*Saturday, 10 September*: Left Baden-Baden about noon and drove to Karlsruhe in an hour – about thirty miles [50 kilometres]. Had a good lunch at my old Schloss Hotel for 2.25 Reichsmarks. Found the same porter I knew in 1928. The town has grown. There are new buildings on the outskirts. Raining heavily. After lunch, I started for Frankfurt on the Richs autobahn. My first experience. It is a simply marvelous road, about thirty feet [9 metres] wide, with one-way traffic and no crossings. Pedestrians and slow driving are not allowed. Drove to Frankfurt in two hours. It used to take four. Spent an hour in Frankfurt, saw Margaretta. Then drove to Cassel for the night. Autobahn most of the way. The autobahn is one of the century's great achievements. All praise to Hitler for it. Cassel is full of Hessians. Hotel good. Garage some distance away.

*Sunday, 11 September*: Got up about 10 a.m. Saw the picture gallery with its marvelous collection of fine Rembrandts and other Dutch masterpieces and most of the other well-known schools. Arrived in Berlin just at dusk. Glad to be here. It is my favorite European city. Herkules Haus, alas, is no longer available. It belongs to the Government. Got two fine rooms in the Adlon for thirty Reichsmarks. Visited Femina and Ciro's. Enchanting as usual! Everyone did the 'Lambeth Walk' at Ciro's. Berlin has the most agreeable night life in Europe, and the prices are amazingly low. One can splurge for a few marks.

*Monday, 12 September*: Up late. War scare!

*Tuesday, 13 September*: Called Teddy in London. It is her birthday. She's 25 years old. Miss her. War scare. Visited Busse. Irmagard, although happily married, is childless.

*Wednesday, 14 September*: Around town. War scare!

*Thursday, 15 September*: Around town. Saw the Kaiser Friedrich Museum. Splendid collection. Saw Lohengrin in the evening – eleven marks. War scare! Talked to Dave Hecht [Getty's attorney] in New York.

*Friday, 16 September*: Lunch at Adlon. Visited around town and then went to Femina for chocolate. Met Phyllis. Didn't look as well as in 1934. Saw *Der Barbier von Seville* in the Staatzoper in the evening. Very good. The best seats are eleven marks. Studied business reports until 3 a.m. Must reform!

*Saturday, 17 September*: Up at 10 a.m. Had three peaches for breakfast. Went to Leipsigerstrasse 76. Talked to Herr Matte of the Reisestelle of the Reichsbank. Found 517 Reichsmarks in my name: 13,444 were transferred in October 1937 to Chase's name. Will transfer them back and save ten shillings per 100 Reichsmarks. According to Chase's record, I drew out approximately 8,500 Reichsmarks in 1936 and 1937. Lunch at Adlon. I had: suppe, Rindelfleisch, gemuse, preisselbeer gemischteseis, ein Kleine helles, trinkel for a total of 4.95 Reichsmarks. I have a salon, spacious double bedroom, a large bathroom at the Adlon . . . forty Reichsmarks daily. Could probably get it for 600 Reichsmarks monthly. Answered Pitt and Scott's wire by letter, saying I will pay duty in New York. Got long cable from Hecht re Hotel Pierre. Must answer. Situation was involved by poor legal work on Gerry's part (*Gerry Realty, the real estate company that owned the Hotel Pierre*).

Order two tickets for *Margarete*, as Germans call *Faust*, at the Staatzoper tonight. Maria Mullen and Roswaenge will sing. To the Femina at five for a cup of chocolate and cake – 1.80 Reichsmarks. There was an amusing crowd, good music, and show. Met Hansi Froshin and Phyllis. To the Staatzoper at eight for *Margarete*. Splendid uncut performance. The Walpurgis Night was most enthralling. Wonderful music! Maris Muller was a mature but lovely-voiced Margarete. Mephisto very good also. The boxes and the first nine rows are only eleven Reichsmarks, compared with seven dollars for the Metropolitan. The Staatzoper seats about 1,900, the Metropolitan 4,000.

After the opera, I had dinner at the Atller, and then a quick visit to the Roxy Bar, which was crowded. Eight Reichsmarks for an Ohio cocktail. Then to Ciro's, three Reichsmarks for an Ohio. Very crowded and gay as usual. Mustafa, the proprietor, is the Sherman Billingsley of Berlin and Ciro's, the Stork Club. Home at 2 a.m. and read for an hour and a half. To bed at 3:30 a.m.

*Sunday, 18 September*: Up at noon. Lunch at the Adlon – four Reichsmarks plus fifty pfennig for a beer. After lunch, I spent one hour reading the Berlin, French and English *New York Herald* newspapers. International situation is very critical. Drove up the Linden to the Schloss, crossed the Kerfuestenbruecke and admired the famous statue. The castle and the morstall are dingier than when I first saw them in June 1913.

*Tuesday, 20 September*: Up at 11 a.m. Beautiful, sunny morning. Breakfasted on four peaches.

*Wednesday, 21 September*: Up at 10:30 a.m. Breakfasted on five peaches. (*Paul seems to have been breakfasting solely on peaches!*)

*Friday, 23 September*: Godesberg! Godesberg Meeting reported a failure. Market weak. Model said get out of Berlin.

*Saturday, 24 September*: Draht called and said, 'Get out of Germany, war is coming!' Beautiful summer weather continues; it is real swimming weather. 'Europa in schpannung' [Europe in suspense] are the newspaper headlines. I saw *Der Troubadour* in the evening. It was a good performance, but the opera drags in places. Roswoenge was good, but is no Caruso!

*Sunday, 25 September*: Up at 10. Talked to Steels of the Amexco. Got 250 Reichsmarks for fifty dollars. Needed the money! Draht called and said, 'Urgent. Get out!' His wife and child are leaving tomorrow. Called Charlotte. Saw the National Gallery upper floor. There are some very good works by Menzel, Thoma Maree and Bocklin. The entrance charge is ten pfennigs. Thirty pfennigs for a führing [guided tour]. Visited the Pergamon Museum again. It is always wonderful. Had lunch with Charlotte. Packed and paid bill. Drove Charlotte home. She got a phone call saying 'no war'. (*Charlotte Susa was a beautiful German actress with connections in high Nazi circles. Her information 'no war' was correct in September 1938. She was a personal friend of Adolf Hitler*). I told her Germany should seek peace. Filled the gas tank, no oil was needed. Last filled at Zurich. Everybody seems apprehensive about war. The Führer is to make a speech tomorrow night. Dinner at Eden Roof! Gay! There were many Spanish-speaking people there, also some Japanese. Paid seven marks for dinner, extras, and tip. Saw *Meimat*. Two Reichsmarks for best seat. Zarah Leander was excellent. She is lovely-looking and a fine contralto. Home to Adlon and read, then bed.

*Wednesday, 28 September*: Beautiful weather. Up at 10:30. Took a long sunbath and ate four good peaches.

*Thursday, 29 September*: Up late. Ate a fruit breakfast. Got started at about 11:30. Just as I was about to drive away, a plain-clothesman sprang into the car, showed a badge, and ordered me to the police station. I was detained half an hour at the station. Thorough search of my personal effects, and the car showed nothing improper. The police were most polite and apologized for the intrusion. They explained times were unusual and promised no further search at the frontier. Had no difficulties at the German frontier, was through in ten minutes. Sent excess marks, about 250, to Amexco Berlin. In Holland had to pay eight gulden in auto tax for fifteen days and had to drive five miles [8 kilometres] to Oldenzaal to change my money. Then I drove leisurely to Amsterdam.

Passed numerous military barriers on the road – felled trees, piles of stones, guards of soldiers. One bridge mined ready to be blown up at any minute. I am amazed at signs of war in Holland. Arrived at the Amstel Hotel in Amsterdam at dinnertime. Found a call from Teddy in London. Talked to her. She was almost hysterical from war scare. However, news from Munich meeting promises peace. Met Model at the hotel. Parked the car outside all night.

*Friday, 30 September*: Up late. Beautiful weather. Lunch at hotel, poor service. Met Teddy and Jean at airport at 5:15 p.m. Am so glad to see them. Back to hotel. We visited Model's in the evening. His wife is a talented sculptress. Must add that I met Myer in the forenoon and discussed furniture, politics, and business. He had a comfortable home. This evening the war scare is over.

*Saturday, 1 October*: Instead of war, peace. Germans march into Czechoslovakia unopposed. Visited Goudstikker. Saw many paintings. Two nice Bouchers 180,000 gulden. One miniature Rembrandt, *Saskia*, 42,000 gulden. Frans Hals 90,000 gulden. Claude Lorraine landscape, 46,000 gulden. Ruisdael medium-sized landscape, 9,000 gulden. Van der Velde still life, 9,000 gulden. I think his prices are no great sacrifice. We had a good lunch at the Astoria for 15 gulden . . . Teddy, Jean, the Models, and I. After lunch I inspected the Mensing collection. There were some wonderful pictures. Afterwards, I drove to Volendam, twenty miles [32 kilometres]. It is a lovely country of polders and dykes.

*Sunday, 2 October* (*having visited the Rijksmuseum at Amsterdam*): We were there only thirty minutes and spent most of it in front of the *Night Watch* – what a glorious picture. It is worth four million dollars.

*Monday, 3 October*: Met Mr Schulein of Texeiro at 3:50. He was very courteous but reserved. We exchanged generalities and then discussed Tidewater. He said he was willing to exchange views if the occasion arose. Schulein is one of the two active partners in Texeiro. The third partner is inactive. Left at 4:15. Met T. at Hentz's. The pound is four dollars, seventy-nine and three-quarter cents. From there we visited Rembrandt's house. It was in private possession until 1906. Now, it is completely restored and contains 250 etchings. Good copies, sold in the old days, fetched one or two gulden. In his studio, sacred holy of holies, shutters he used to regulate light are still there. Brockhurst [Gerald Brockhurst, RA] has a better studio.

*Tuesday, 4 October*: Met Myer at 11:45, and we discussed T.V. He will be in New York at the end of October, to remain some months. Promised to sound out the situation here and discuss it further then. He would like to move to America.

He says Europe will have a big war in two or three years, that England will not tolerate much more. Lunch at hotel, then got the car. Visited Goudstikker and saw a Carlin console, which is really a dining room piece, with five Sèvres plaques – four narrow ones, one large, with basket of flowers. White marble top (cracked), white marble center jambes, 'last price' 30,000 gulden. It is being sent to America. Believe it is worth 20,000 g. as a bargain. 30,000 g. is dealer's price to public. I don't think it will bring over 2,000 guineas at Christies.

*Wednesday, 5 October*: Visited the Mauritshuis Museum. All my attention and admiration for Dutch 17th-century art. The car was returned, bill 12 gulden. Drove to the hotel through fierce rain. Lunch. The Crown Prince (of Kapurthala's) Rolls is waiting. Major Dass is paying his own and his master's bills as I pay mine! I spoke to Ronny at Lugano and then left in driving rain for Rotterdam. Passed picturesque Delft.

(*On the same day, he visited the Boymans Museum*): Saw special exhibition of four masters. Some magnificent 17th-century Dutch pictures. Some miserable impressionistic moderns. *Marten Looten*, a splendid portrait by the young Rembrandt, impressed me. Left for Antwerp. (*This is an interesting entry because he bought the painting of Marten Looten by Rembrandt.*)

*Saturday, 8 October* (*in Paris*): Visited the Fabre collection. The furniture is not as sensational as I had expected, but in Savonnerie, he bowled me over. It's the greatest collection, in quality and quantity, that I have ever seen. There are two magnificent Louis quatorze Savonneries from Santiago de Compostela, which had been given by Louis XIV to the Duke de Choiseul, French Ambassador to Madrid, and by him to the Archbishop of Toledo. Dimensions and prices are: 8.90 metres by 4.60 [29.2 feet by 15] – $65,000, 8.90 by 3.30 metres [29.2 by 10.8 feet] – $45,000. They are in perfect condition and no repairs are necessary. Saw a bureau à Cylindre in mahogany, with simple ormolu signed 'Riesener', 220,000 francs. Also a better one from the Rasmussen sale for 220,000 francs. However, my Molitor is worth double. Saw the Bauer collection. A tric-trac and card table was interesting, price 150,000 francs. The rest of the collection did not interest me. Visited Louvre. It is still ninety percent closed on account of the late war scare. Talked to New York re Pierre [Hotel]. Visited Ball at Jansens in the Rue Royale. Saw his Louis quatorze Savonnerie again. I've offered fifteen hundred pounds plus five percent commission for it. It is 5.90 metres [19.4 feet] long, 4 metres [13 feet] wide, and is in perfect condition, but both ends are cut off. Dinner with Teddy, Jean, Kathleen, and Chatin at the Escargot. Then we went to the Bagdad and the Boeuf sur le toit. 25 to 35 francs for each short drink, 110 francs for a quart bottle of P and G 1928 nature (*presumably Pommery & Greno 1928*) at the Escargot. Boeuf crowd was disturbed by a drunk playing with a fire extinguisher.

*Monday, 10 October*: Talked to Fini at Lugano. Ronny's well. La Croix came. We visited Ball and saw a magnificent secretaire, three plaques, pate tendre, very rich in ormolu, priced at only eighteen hundred pounds. La Croix dated it at 1800 to 1810. I am sure it is 1785 to 1790. It was sold at auction in 1825 for 6,200 francs to the Duke de Cambaceres and was bought by Ball from the family for fourteen hundred pounds. I offered fifteen hundred pounds and at 4 p.m. a letter accepting my offer was received. Had dinner at the Ritz, 75 francs. Read in the evening. Teddy has already left for London. I was lonely when T left!

*Tuesday, 11 October*: Ball telephoned that he is ready to show another Savonnerie. I had my shoes soled and new rubber heels put on for 110 francs. Following are some Paris prices: Tout Paris – crème de menthe, 40 francs, no cover charge; Bagdad – sandwich, 15 francs; gin fizz, 35 francs; Johnnie Walker and soda, 45 francs.

*Wednesday, 12 October*: Fini arrived to talk about her father's case. I advised a settlement. (*Getty's former father-in-law, Fini Helmle's father, was arrested by the Nazis, in order to extort money from him.*)

Getty's entries at this time are almost exclusively concerned with seeing beautiful pieces of furniture or carpets, noting the prices, and in many cases, making offers that are accepted. On Thursday, 27 October, we find a particularly interesting entry: 'Up at 11. Charlotte [Susa] telephoned regarding the furniture. Met her for lunch at the Adlon. She reported a new law re the Jews and that Rothschild's furniture may be sold in January. She will keep me posted.' It must be remembered that Europeans and Americans were quite unaware of the stark horror of the fate of the Jews in Germany at this time and that in this case, two Rothschild brothers were being blackmailed by the Nazis to ransom the third.

*Thursday, 3 November*: Up at ten. George F. Getty, Inc. owes brokers nothing, Chase 750,000 dollars. No immediate prospect of reducing this loan. Regret it. P.W. owes Chicago Bank $2,800,000 and the Chase 150,000, which will soon be paid off. the $2,800,000 is a heavy burden with its $70,000 annual payment plus $98,000 interest. Total compares with $650,000 interest before K.J. sale. Estimate P.W. will have to sell 150,000 T.V.N. and G.F.G. 50,000. T.V.N. to pay off bank debt, leaving P.W. 145,000 T.V.N.G.F.G. 350,000 T.V.N. Spoke to Ogilvy at the Embassy, and I am about ready to leave. Paid hotel bill.

*Saturday, 12 November (back in New York and meeting with Lord Duveen)*: Met Lord Duveen. He is 69 years old, and he looks 55. Also saw his nephew, Lowengard. Duveen is a very interesting, very fine type of man. He held me enthralled by his conversation until 7:20 p.m. Saw my carpet [the Ardabil carpet], very pleased. Noticed it is intact, except for modern border on one end and four foot long piece on border at left side. It was bought in 1919 by Duveen for 57,000 dollars. At Anderson galleries, Delamar sale. Duveen was on a boat at the time, and cabled his firm to go to 200,000 dollars if necessary. My carpet is about 24 feet by 12 feet [7 metres by 3.5 metres]. Duveen has another slightly larger carpet for which he paid the Duke of Anhalt 100,000 dollars plus ten percent. The carpet belonged to the Sultan of Turkey and was taken in the spoil in Vienna in 1683. It's not equal to the Ardabil. He told me he sold Mellon $12,500,000 worth of art just a few weeks before Mellon died. The total sales to him were $35,000,000. He sold Frick $7,000,000 in one year. He sold Fragonard to Frick for $1,250,000 after buying it from Morgan Junior at the same price. Would give two million for it today.

Discussed Hillingdon collection. Most important piece of Sèvres furniture in the world is the DuBarry commode, bought forty years ago by Paris Rothschild for $60,000 from the Count Cheremetiev. One Hillingdon piece is the second most important. (*Prince Cheremetiev appears in Paul's discussions on his journey to Russia in 1935, pages 95–6.*)

1939

*Monday, 10 April*: At office in afternoon, Pierre decision is discouraging. Caused by failure to set up arrears in rent correctly in October and by not joining proper plaintiffs. Received a letter from Samuels. Bought all four rugs catalogue numbers 67, 69, 70, 71 for $9,922, including five percent commission. About twenty-three percent former value. 67: Indo-Ispahan hunting carpet, 2000 dollars, about 1625 in date, 6'6" × 4'7" [1.98 × 1.4 metres], sold at William Solomon sale in 1923 for 9,000 dollars. 69: Polonaise, 3,900 dollars, about 1600, 7'0" × 4'10" [2.13 × 1.47 metres] sold at Yeakes sale, 1910, for 12,300 dollars. 70: Ispahan, 1,200 dollars, late 16th century, 15'5" × 7'0" [4.7 × 2.13 metres], sold at Clarke's sale in 1925 for 8,000 dollars. 71: Polonaise, 2,350 dollars, 1600 in date, 13'7" × 5'9" [4.14 × 1.75 metres], sold at Benguiat sale 1925 for 13,500 dollars.

*Friday, 14 April*: Had wire from Hecht. Chandler act proceedings against Pierre Hotel dismissed, but possession was not given. 2 E. 61st Street Corporation will file voluntary bankruptcy Monday. Glass of milk and crackers for dinner. Then at 8 p.m. attended the Jewish synagogue for the first time. A large crowd was there. Eddie Cantor spoke. I suppose I was the only Gentile in the place!

*Monday, 28 August (now back in Europe)*: Decided to concentrate on Greco-Roman things as I like them best; and an occasional painting new or old. Negotiations between Germany and England. No war yet. Jean buys thirty litres of gasoline and fills tank of her car. After midnight, gasoline will be rationed. No car can leave Switzerland with more than ten litres of gasoline.

*Wednesday, 30 August*: I still hope for peace, but I have believed since I told T and J to leave Rome that war is likely.

*Thursday, 31 August*: Listened to the radio news at 10 p.m. from Berlin. Looks like war! 16 points rejected.

*Friday, 1 September*: Slept quite well last night and felt better this morning. At 10:30 a.m. heard Hitler speak in historic session of the Reichstag. Heard law passed re Danzig. Means war. Simply awful! Felt anxious about T and J! Stuttgart radio news at 10 p.m. gave details of German advance in Poland. Switzerland is mobilizing. Everything is topsy turvey!

*Sunday, 3 September*: Phoned Charlotte in Berlin at 11:30. Quick and good connection. Impossible to phone France or England. At 12:30 got news of England's Declaration of War against Germany from the Swiss Radio News.

*Tuesday, 5 September*: J drove T and me to Interlaken. Took Lotschberg Railway at 3:45 for Milan. Had no trouble at frontier. No Swiss inspection. Italians very kind and courteous.

*Friday, 8 September*: Met Vangelli at 10. Pick up J and drove to see my bust in the workshop near the Vatican. Suggested some corrections to it. It is forty days' work to carve the bust in marble with modern tools, including electric drills. In the old days, it took five months. It will be ready in ten days. Crestola marble seems very white.

*Wednesday, 13 September*: Dinner at Ambassadors. Teddy's hair caught fire while she was blowing out the candles – scared me! I quickly put it out with my hands – no damage.

*(On 25 September he returned to Germany, which was at war, and sees a good deal of Charlotte Susa.)*

*Friday, 6 October*: Lunch at Adlon with Charlotte. Walked in the afternoon and saw Hitler, Göring, Hess, Ribbentrop, Goebbels on their way to the Reichstag. Heard Hitler's speech and thought it worthy of consideration. Met Charlotte at 7. Walked with Charlotte in utter darkness along the Linden after the Opera.

*(On Sunday, 8 October, he returned to Rome and resumed his interest in buying antique marbles.)*

*Tuesday, 14 November*: Married at noon in a romantic setting. A palatial room in the Campidogilo. Spent afternoon packing. Dinner with Teddy. Left for Naples at 8:30. T and J came to the station to see me off. Much disappointed that T decided not to go.

*Wednesday, 15 November (in Pompeii)*: Hurry to catch the Conti de Savoia due to sail at 12!

Paul Getty had made his last attempt at marriage, this time to Teddy Lynch; but a marriage in which the first two and a half years are spent apart cannot be said to have started under very auspicious circumstances. Then, as Getty explains in the tapes (on page 71), when she did join him in Tulsa, he was so heavily preoccupied in the aircraft business for the war effort in America that he found very little opportunity to console her. So, when she received an offer to go and sing in Hollywood, she could not refuse and, naturally, they began to drift apart.

They did not officially separate, however, until 1951, and the divorce did not come until 1958. Therefore, when Getty reappears in Europe in 1951 and settles in a house in England in 1960, it is as a bachelor . . . and a bachelor he remained.

## Letters Getty Treasured

*From Karen de Woolf Grady dated 21 November 1957*

Dear Paul

Reading about you made me wonder if you knew how successful you've made others also. I, for one, owe everything to you, not only because you lent me money when I was desperate, but because of the way in which you did it. That alone enabled me to write my first novel. Bobbs Merrill brought it out, and whilst in itself, it stirred tremendous disinterest, it did get me an M.G.M. contract. Since then, as Karen de Woolf, I've always drawn top salary as a writer, and after the death of my husband, I was easily able to raise my children and properly care for my parents – all because of a day during the Depression when, in answer to my urgent call, you made the time to meet me.

I was preparing to foal and felt awkward and ugly, but you pretended I was attractive – even said something flattering. I don't recall what. It did more for me than the check. I'd never asked anyone for a loan before, but despite my agony of embarrassment, my husband had to have help. You acted as if the use of your money was the most natural thing in the world. By the time you were paid, I was doing all right, but the more important thing was, what I had learned from you, Paul. Since then, I've had opportunities to help others, and I've never forgotten the pains you took to put ME at ease.

The reason I turned to you was because you'd been so wonderfully kind to me when I first came to Hollywood. I was Gypsy Wells then, intending to allow the picture business to improve its product by using my story material, meanwhile writing poetry that didn't sell and convinced that all men had a fate worse than death in mind for me. Why, I can't imagine. I certainly was no Marilyn Monroe, nor as I see it now, did I even seem to have all my buttons! But you were the guy who never made a pass – just found time to haul luggage when I moved, bring ice cream when my throat was sore, and take me for drives, which I loved.

After any time spent with you, I was always comforted and reassured, especially on one occasion when you should have had a medal for gallantry. Being in my 'teens you were, to me then, an older man, and hoping to look like the kind of woman I fancied you squired, I had borrowed a fur jacket from the girl next door. Obviously, she hadn't used it for some time – but moths had. We weren't far in your open car when the fur, literally, began to fly. I used the situation in a script once, and it was hilarious, but it wasn't to me that day. If you'd laughed, even mentioned the matter, I should probably have flung myself from the car in disgrace. But you just went on making comfortable conversation and picking horrid hairs from your mouth and lashes as if fur in your face was normal activity, until I took the little monster off and sat on it. I adored you for that, but I won't embarrass you by an eulogy.

I just want you to know my deep appreciation for the sense of safety you gave a scared kid, and the gracious way in which you helped me as a married woman. It was knowing that I had such a friend to turn to that made me feel secure enough so that I never had to ask for help again. My son and daughter are both married now, well and happily, and since I'm thirty year after year, it won't be long before they pass me. But it wasn't until they were in college that I married again – my attorney, Dennis Grady, and although it's certainly no longer necessary that I work, I enjoy it – television only though, mainly originals on film, which I do at home in my own time.

So, my life is full and, to me, entirely satisfactory. And you're the one who made it possible. Knowing your kindness, you must have opened similar avenues for many others, so naturally you've succeeded spectacularly – a feat so nearly impossible in this day and age that it must have been tremendous sport. And whilst it's highly improbable that you'd ever need a loan from me, it's not impossible that you could use my help in some way. When that time comes, call.

Karen de Woolf Grady

*Getty on Getty*

From Donald K. Phillips dated 19 November 1956

Dear Paul

When Dave talked with me on the telephone, I was feeling so utterly miserable and the world had such a bilious green look about it that I honestly wondered if I had accepted – if I had accepted whether I could have been around to attend the first meeting. These factors prompted my regretting today that I could not accept. Now that I feel better, I realize that I made a very bad decision, for it could only have meant an opportunity to learn much and improve my stature had I accepted your invitation and been approved a Director of your companies, Skelley, Tidewater & Mission Corporation. If in the future, the time ever arises when you're seeking a replacement for a retiring Director and you feel that I might qualify . . .

I look back with very fond and happy memories on the wonderful education I got when you were actively in the process of building up your empire. You were nice enough to trust me with some of your very important orders. If I had just been smart enough to have taken at full value your statement that Tidewater, which was then selling at three to three-and-a-half, would one day be worth a hundred and twenty-five – you know what the present stock is worth in relation to the old – that Skelley should be worth somewhere in the neighborhood of two hundred dollars a share – despite the fact that it was then selling, as I recall it, in the 'teens – and that Mission Corporation which, if I remember, we bought in hundreds of thousands of shares at nine to sixteen, would one day command a price of seventy-five to a hundred, how simple it would have been for me to be a relatively rich man today by investing only a few thousand dollars, which incidentally, I didn't have. It's fun to think about, though, isn't it?

Seriously, Paul, I think what you have accomplished and the fact that you were willing to predict – over twenty years ago – exactly what was going to happen, is one of the most amazing stories of financial success that I've ever heard of.

I've tried a couple of times since George took over the seventeen Battery Place operation to get over to meet him personally, but he's been – on each occasion – either very busy or on his way to a meeting or at a meeting, and I still haven't had the pleasure of shaking him by the hand. But from everyone who has come into contact with him, I hear only one report, and that is that he's a man of tremendous capacity, ability, and personality, and a real chip off the old block. I'm sure you must feel a great sense of pride in his accomplishments.

This has proven to be a regular book, hasn't it, Paul, but it has been much too long since I have talked with you or corresponded with you, and I enjoyed visiting with you and hope you may get a nostalgic kick out of thinking back to the old days when we used to buy a few yachts and transmit a few kronen to Copenhagen, as well as accumulate Mission, Tidewater, and Skelley, to say nothing of a few hundred thousand shares of Petroleum Corporation of America!

I'm off on the doctor's orders for a period of recuperation in the South, where I can soak up some sunshine, but when I return to the office, which will probably be around the first of the year, I'll give you a ring on the phone and have a man to man visit.

Always my warmest regards and best wishes for a holiday season filled with every good thing. Sincerely,

Your friend, Don

# 12
# RECOLLECTIONS OF A LUNCHEON

On Wednesday, 7 November 1973, my future wife, Juliet (Marchioness of Bristol), and I were entertained by Paul Getty at lunch. The occasion was to round off the series of conversations we had been having. Lady Juliet and I arrived at one o'clock, and Paul joined us in the drawing-room, sitting in the same chair I had always used when interviewing him.

We began by discussing *The Waterloo Campaign*, a book he had been reading (published by the Folio Society, 1957), and Paul asked my opinion of Marshal Grouchy, Napoleon's marshal.

'Grouchy's instructions were to follow Marshal Blucher, Wellington's allied general', I recalled. 'He was to bring Blucher to battle but Blucher got away from him. One of Grouchy's generals came to him and said that he had served with the Emperor during the Italian campaign and added: "I have heard the general enunciate a hundred times that you should move towards the sound of guns. The Emperor is at grips with the enemy at this moment, and we should proceed towards the battlefield."

'But Grouchy stuck to his orders as he had read them and continued his vain search for Marshal Blucher. Napoleon in his memoirs reproached him for not having gone towards the cannonade of Waterloo and stressed the slowness of Grouchy's movements, to the extent of referring to the fact that they were waiting for soup to cook, and so forth.'

At this point, Paul observed that 'there was no real reason why Grouchy should not have moved as fast as Blucher. The Germans were men, just as were the French, and presumably Blucher was able to move fast enough!'

Then Paul launched into the subject of underlings: 'One is very nearly always let down by underlings. They may be all right for 80 per cent of the time, but for 20 per cent, they do something quite incredible. If you ask them to show caution, they go and do something reckless; if you ask them to be bold, they show extreme caution. This is what makes it so difficult for leaders and great men, this relying entirely upon subordinates.'

> **On the subordinates**
> One is very nearly always let down by underlings. They may be all right for 80 per cent of the time, but for 20 per cent they do something quite incredible. If you ask them to show caution, they go and do something reckless; if you ask them to be bold, they show extreme caution. This is what makes it so difficult for leaders and great men, this relying entirely upon subordinates.... Sometimes you get enthusiastic about somebody and then he does something you feel was a real boner.

It occurred to me that he was speaking very much from his own personal experience. He went on to refer to Napoleon's failing judgement during the five years prior to the Battle of Waterloo and, of course, quoted the case of the Russian campaign.

'I sometimes worry whether my own judgement can be failing,' he said. 'For example, I feel that I'm running scared, because one cannot be certain that one is making decisions with the same certainty of touch.'

> **On judgement**
> I sometimes worry whether my own judgement can be failing. For example, I feel that I'm running scared, because one cannot be certain that one is making decisions with the same certainty of touch.

Paul referred to President Nixon, stating that he thought that Nixon showed a lack of judgement. Paul mentioned that if the tapes were to be handed over at all, they should have been handed over earlier. I found this reference interesting in light of his previous statements (see pages 117–18), when he had defended Nixon fairly vigorously and said that it would have been very wrong of him to hand over the tapes of private conversations due to the fact that it would compromise the privacy of all presidential conversations.

We went to lunch. The table in the dining-room was an immensely long refectory table that Paul said had belonged to Randolph Hearst. 'Hearst and I shared a liking for long tables,' he said. Indeed, the one at Sutton was extended by a second table, and they stretched nearly the length of the room. Paul sat at the end near the door with Juliet on his right and me on his left. We were sitting in Charles II chairs.

Lunch was private, quiet, and served by a butler – very pleasant. Some extremely good claret was sipped. Port followed. I noticed that Paul, who drank comparatively little, did drink a small amount of white wine, which was served with the first course (a very good hock Rhine wine) and a glass of port.

During lunch we talked to Juliet about horseracing as she had a filly called Coralivia in training and her mother had various racehorses. Paul said his son once had a racehorse running at Ascot. Paul had been somewhat surprised when it was handicapped 14 pounds (6.5 kilograms) heavier than a horse that had run third in the Derby. Juliet pointed out that the handicap was due no doubt to the other horse never having won another race, whereas his son's, George's, had won a couple of smaller races in Ireland. Paul was not very convinced by this logic. He agreed with me that racing is a hazardous and unsatisfactory business.

Horses remained the subject as Paul referred to the 'King horses', as he called them. In particular, he discussed a horse ridden by a Mexican bandit, Joaquin Murietta.

'Murietta was being chased by the American army and came to an enormous chasm. He managed to jump the chasm with his superb horse, but the horses and riders with him balked. He could have got away due to his superior horse, but a sharpshooter in the army brought him and his horse down.'

We all felt rather said after the horse and rider in Paul's story had made such a gallant leap.

'There is a place called Murietta's Leap,' he continued. 'On another occasion, there were some customers in a dance hall in San Francisco who had gathered together a pile of gold on a table. They were boasting about what they would do to Joaquin Murietta if they caught him, as there was a large price on his head. It so happened that Murietta was in the room. He came forward and said: "I'm Joaquin Murietta. Just exactly what would you do?" At that moment the lights went out. When they came on again, there was no sign of Murietta or the gold.' Paul chuckled a good deal at the story.

The subject changed rather markedly to weight-lifting. Paul said that the heaviest weight he had succeeded in lifting from the ground was 230 pounds (104 kilograms). I thought this a colossal weight and asked him why, in particular, he had chosen weight-lifting as a hobby.

> **On weight-lifting**
> You can't cheat yourself. In other sports you can always ease up a bit or say you weren't feeling well or that the other fellow was too good and so on. But when it comes to lifting weights, you can't cheat. Say you succeed in raising 200 pounds (90 kilograms) into the air, then if you add another 5 pounds (2.3 kilograms) to that, you've got to make an even greater effort. It is, in fact, the final test of your endurance.

'Well, you can't cheat yourself', he replied. 'In other sports you can always ease up a bit or say you weren't feeling well or that the other fellow was too good, and so on. But when it comes to lifting weights, you can't cheat. Say you succeed in raising 200 pounds [90 kilograms] into the air, then if you add another 5 pounds [2.3 kilograms] to that, you've got to make an even greater effort. It is, in fact, the final test of your endurance.'

I asked him how long it had been since he had lifted weight of that kind. He answered: 'It must be forty years.'

Getting off the topic of weight-lifting, I asked Paul if he knew Heini Thyssen well. He said Heini was a friend of his, and I told him a story Heini had told me about the time he had invited Paul to the Villa Favorita on Lake Lugano. The villa got so full of Arab oil sheikhs and others coming to see Paul about his business deals that eventually Heini had to move out himself! I said to Heini: 'Why don't you move into Sutton?' Paul laughed a lot at this and said that the story was true.

After lunch Paul showed us around the house. When we came to the throne of Juliano de Medici, on which the prince could be flanked by two admirers, he recounted an amusing story.

> **On lawyers**
> A story my father used to tell me, about a rich man who lay dying, and sent for his two layers. They came, rubbing their hands. He said to one of them: 'Stand on this side of the bed.' To the other he said: 'Stand on the other side of the bed. . . . And now, like the blessed Lord Jesus, I lay me down to sleep between two thieves.'

'It reminds me of a story my father used to tell me about a rich man who lay dying, and sent for his two lawyers. They came, rubbing their hands. He said to one of them: "Stand on this side of the bed." To the other, he said: "Stand on the other side of the bed." They just stood there, hoping for his blessing. And he said: "And now, like the blessed Lord Jesus, I lay me down to sleep between two thieves." '

Before he left, Paul took me through to his study to find some letters he wanted me to see for the book and the photograph of the Duchess of Carcaci. She was Marie Regina Paterno Castilla, née Millington-Drake, born in Brussels on 19 September 1924. The picture showed an attractive brunette in her lace bridal dress and veil with four ropes of pearls. On the back were printed extracts from some holy works, including: 'I beseech all those who love me to grant me the help of their prayers' (St Ephraim). He seemed to attach importance to the photograph, which was mailed to him after her death at Taormina on 25 July 1973.

I believe Paul had wanted to marry her before she became

the Duchess of Carcaci, but was disconcerted to discover that her parents had been having him checked. He thought: 'Oh, my God. Another fortune-hunter.' What he did not know was that her parents were extremely rich and were having Paul checked because they had never heard of him and thought he was after her money!

It turned half past three, and we offered to leave. As usual, Paul showed us to the door. We left the hospitality of Sutton and Paul Getty ever more aware of him as a warm and excellent host.

# 13
# CONCLUSION

In the Introduction I stated that the problem in approaching the life of Paul Getty is that the whole is much greater than the parts. I was not trying really to contradict the laws of nature because there was, in fact, a missing element in Getty's character not immediately apparent – genius. One is reminded of the analogy of Madame Curie, who, having reduced the elements of a certain substance to all its constituent parts, found that there was still one ingredient missing. It was not until she discovered that mysterious ingredient glowing in the dark that she identified it as radium. The mysterious glowing radiation, invisible in normal circumstances in Getty's personality, was, indeed 'genius'. It was the secret of his magic. The outlines of his life on paper appear much like those of other successful businessmen, but his innate genius, combined with a duty to detail, made him 'A Man in a Billion'.

Getty told us on the tapes of his acquisitive nature when a child. He also related how on his twelfth birthday he was selling copies of the *Saturday Evening Post* in order to swell his fortune to $12 – despite the fact that his family was wealthy. We find him joining his father in the oil business at an early age, where he shared ardours and dangers as a working man on the rigs. We discover that he made his first million at the age of twenty-three and retired to become a playboy. That could have marked the end of his talent for business and his rise to corporate fame. However, he tired of the playboy life very quickly and resumed the life of an active businessman. I do not think Getty would have known how to lose money. If knowledge or flair did not come to his aid, luck did.

Getty interests poured over $20 million into the Arabian oil gamble before a drop of oil was found. The project to start

## Conclusion

explorations for oil in the Neutral Zone between Saudi Arabia and Kuwait was not only uncertain, but enormously difficult. Pipelines, the refinery of Mina Saud, settlements for employees – all had to be built in the most inhospitable climate in the world and thousands of miles from sources of supply. As luck and foresight would have it, he had been quick to realize the potentialities of the house trailer market in the USA and, after the Second World War, had switched from the production of aircraft at Tulsa to the production of trailer homes. Thus, he had a ready-made supply of homes to transport to the Red Sea complete with refrigeration and sanitary arrangements.

Three years of waiting and hoping were to go by with mounting costs before the initial strike in 1953, right in the middle of the concession. A test well drilled to 3500 feet (1100 metres) reached the Burgan Sand. In all, 7 million barrels of crude oil were produced there during the first full year of operation. Next, a test drill to 7000 feet (2100 metres) tapped another oil horizon, the New Ratawi Limestone. Two drills there yielded more than 5 million barrels a year. Then came three additional Eocene limestone reservoirs at different levels ranging from 500 feet to 2000 feet (150 to 600 metres) below the ground. The whole area, he once said, turned out to be like an enormous layer cake, with numerous separate strata and great reservoirs of petroleum sandwiched between the rock and soil. According to impartial geologists, the area was shown to have at a conservative estimate 13½ billion barrels of proved crude oil reserves in place. Getty referred with gleeful relish to the prophets of doom on Wall Street who thought he had met his Waterloo on the sands of Arabia's deserts.

Yet, his greatest victories may be seen in the realm of art. It may well be that in the purchase of the Ardabil carpet of AD 1515 (valued now at $20 million and hung in the Los Angeles Museum) he achieved a greater triumph than in gaining control of Tidewater Oil.

In his book, *The Joys of Collecting,* Paul wrote: 'In my opinion an individual without any love of the Arts cannot be considered completely civilized.' He goes on to give an extremely interesting account of how he acquired one of the most important paintings in his collection at an auction in Sotheby's in 1938 for approximately $200 – *The Madonna of Loreto* from the collection of the late Jaimee II of Bourbon. As he told it:

I decided I needed some expert advice before bidding on The Madonna of Loreto and the Louis XIV portrait. Gerald Brockhurst, the well-known English portraitist, who, I might add, painted my own portrait that same year, acted as my adviser in regard to the Madonna. He recommended that I purchase the panel for he strongly suspected that it was not simply an 'after-Raphael' painting; he believed the foreshortening of the Virgin's right arm betrayed the master's own touch.

Leon Lacroiz, an expert on French eighteenth-century art, gave me his opinion on the Louis XIV portrait. He thought it a good example of Rigaud's work.

And so I decided to buy, if I could, both these paintings. I really didn't have any set price limit in mind. I think that nearly all the art dealers of London and Paris, as well as numerous museum experts and private collectors, were present at the sale, and I expected stiff competition in the bidding.

As the sale progressed, it became apparent that those present were not inclined to pay high prices. When the flower pieces attributed to Van Huysum were auctioned, I bought it with a top bid of 55 pounds, in those days approximately $275. A few minutes later, the Madonna came up for sale. It was a panel that seemed unprepossessing at first glance. In fact, it was in somewhat poor condition. It dealt with a classic subject, a portrait of the Holy Family. No special claims were made for it. The painting, it seemed, was a copy of Raphael's famed, long-lost Madonna of Loreto. It might have been executed by one of Raphael's students or contemporaries. Apparently, no one else attending the sale had anything approaching my interest in the panel, and evidently, none had seen what Gerald Brockhurst had noticed in the painting.

I waited for the opening bid, which proved to be ten pounds, or about $50. By then an experienced and cautious auction buyer, I increased the bid only slightly, which someone promptly topped by another few pounds. And so it went back and forth, until I bid 40 pounds – roughly $200 – and there was no further competition. Lot

## Conclusion

No. 49 was mine. My 40-pound bid was more than anyone else was willing to pay.

My luck held when Lot No. 136 was offered. I purchased Rigaud's portrait of Louis XIV with a winning bid of 145 pounds (about $725). Soon afterward, I had the items I had purchased at various auctions shipped to New York.

My liking for the 'after-Raphael' panel continued to grow after it was in my possession. I was constantly drawn to it, intrigued by it. And, I had confidence in it. To employ a colloquialism, the painting 'had something' – something which set it apart, a quality which exerted an ever-increasing attraction and fascination.

Years – twenty-five to be exact – passed. Although my collection expanded greatly, and I had been lucky enough to acquire several very important additions to it in the interim, the $200 Madonna remained one of my favorites. In 1963, I had the painting shipped to Sutton Place.

There were some abrasions of the original paint, and there seemed to be thick daubs of repaint and discolored varnish covering the panel. A few days after the painting arrived at Sutton Place, Colin Agnew, the prominent London art dealer and expert in Italian Renaissance paintings, visited Sutton Place, and I asked him to look at the Madonna.

He was not impressed by the work. Quite the contrary. He asked a friend of mine why I had purchased it. When told that I thought it was a Raphael, Colin blinked.

'Who in the world ever sold that thing to Paul as a Raphael!' he exclaimed.

Colin did suggest, however, that the painting needed cleaning badly. I followed his advice and sent the panel to him to be cleaned, and when the work was finished, the quality of the picture became evident. At this point, Colin became convinced that the Madonna of Loreto was by the master's own hand.

Subsequently, on the advice of a leading art historian, I had the picture given a thorough 'stripping'. This involved the removal of all or almost all of the repaint to reveal the original work. As a result, the great quality of the painting became evident. Now one could see definitely the early sixteenth century work of the master, Raphael, free of the disfiguring repaint of later generations.

Colin Agnew consulted with Dr. Alfred Scharf, an authority on fifteenth and sixteenth century Italian paintings. After meticulous study of the panel, Dr. Scharf accepted it as the autograph work of Raphael.

The discovery caused a considerable stir throughout the art world. Other experts who have since examined the panel and the infra-red photographs and X-rays taken of it concurred with Dr. Scharf's verdict.

As these words are being written, the Madonna of Loreto is on display in the Raphael Room of the National Gallery in London, to which great museum I loaned the painting in February 1965. It hangs next to another masterpiece of the important Raphael – his Aldobrandini Madonna.

Insofar as the monetary value of the Madonna of Loreto is concerned, authenticated as a genuine Raphael, it is virtually priceless. It is insured today for 10,000 times the price I originally paid for it.

I regret to say that, since Paul's death, doubt has been cast on the authenticity of this picture since another version, thought to be the original, has come to light.

In spite of all his successes, Getty did not think he had the Midas touch. He knew how hard he had to work on his projects. As Napoleon said: 'Everyone thinks that power came to me as if it were of itself. But I know what pains and vigils and arrangings it has cost me.' Getty would not have touched anything unless he knew he could turn it into gold – by hard work.

According to Ralph Hewins in his admirable book on Getty, *The Richest American* (published by Sidgwick & Jackson in 1961), the original John Getty had come to America from Ireland by way of France. He opened a tavern on National Road in a place called Cresaptown, Allegheny County, Pennsylvania, in 1817. He had married in France in 1791 or 1792, but presumably his wife died, because late in life he married again. He was said to have become an alcoholic and in about 1830 fell off a horse and froze to death. This would have been Getty's first direct American ancestor.

Perhaps as a result of the fate which had overtaken this ancestor, the publican, the family was noted for its abstemiousness, and this persisted in Getty's case. He drank very sparingly, although he always made sure that his guests had very good vintages.

One of his family, George Getty, was a general in the Civil War. His picture shows a fine-looking man of about forty with dark hair, moustache and beard and piercing eyes. It is labelled: 'George W. Getty led a division in the army of the Potomac.'

However, the original Getty's grandson, William Reid Getty, was really the first member of the family to achieve great public distinction. He was elected Justice of the Peace thre times beginning in 1859. In 1864, after having been in the leather business for a while, he went into politics in Grantsville. He was a Democrat of unimpeachable integrity and became a senator.

Getty's own father had begun life having to take over the responsibility of looking after his family at the age of six. After college, he went into law and then branched out into the oil business.

Of course, Paul was always careful with money – you do not have the talent for acquiring $2 billion otherwise. But his face was as hard as a lemon when he said to me in the last recording we made: 'It is hard to make money; it is very difficult.' This is not much consolation for those readers who expected to find this book an easy formula for getting rich without really trying. It did pay Paul, however, to gamble on his instincts. In his character he possessed seemingly contradictory qualities: one an extreme, almost excessive, caution, and the other, once he had made up his mind, gambling on a seemingly reckless scale.

During the Second World War, while bombs were falling and battles were being fought in frail aircraft overhead, Getty was busy turning the Spartan Aircraft Company of Tulsa into an efficient war-making machine. He worked fourteen to sixteen hours a day and was entitled to a holiday. From November 1940 to May 1941 he took a holiday with Hal Seymour to Acapulco, Mexico. There he heard of a fabulous beach called Revolcadero. It was almost impossible to approach by land as it lay at the end of 15 miles (24 kilometres) of jungle. Paul, remember, always was an avid beachcomber. While all his men stood silent on a peak in Darien, he saw beside the shark-infested waters the beach of his dreams.

In spite of the impossibility of owning the land under Mexican law, in spite of the lack of landward approaches, in spite of the sand that could blow into the bungalows, he overcame all the obstacles and established the Pierre Marques Hotel there. It became a sort of Paradise. In 1971 he was able to unload the whole thing to a hotelier further down the coast for an overall profit of $6 million dollars. He just could not lose.

After this it seems almost superfluous to mention that he bought the most fashionable hotel in New York on Fifth avenue – The Pierre – for $2,350,000. After improving its administration, he sold it in 1958 for ten times what he paid for it. But here again we find that it is not just a case of shrewdness in buying a valuable property cheaply, but that he devotes himself meticulously to the hotel business, studying it in Switzerland, France, and elsewhere in order to be able to improve the property and turn it into an enormous success.

The highlight of his business career would seem to be his struggle to obtain the control of Tidewater Oil, a vast company with refining capacity that he wanted to ally with the production of the crude oil George F. Getty and Company possessed. This came about eventually through his securing the Rockefeller holdings in Mission Corporation, which ultimately controlled Tidewater Oil.

Two important ingredients in Paul Getty's success were his charm and self-confidence. They are best illustrated by the way he extracted himself from an awkward situation during his early oil days. Most of the big refineries were trying to strangle him by denying him refinery capacity. His stocks of crude oil were piling up with no outlets, and storage was becoming a critical problem. He heard that Sir John Legh-Jones, head of the Shell Oil Company, was visiting Los Angeles. Although only a young man and, at the moment, an unpopular wildcat, he boldly asked for an interview. 'If you are going to get "no" for an answer,' he thought, 'you might

## Conclusion

as well get it from the top as from the bottom.' He impressed Sir John, who was shocked to learn of the tactics of Paul's rivals. He immediately offered Paul the backing of Shell in relieving him of his embarrassing stocks of crude oil. Once more, Paul Getty survived by a narrow margin and was to leap forward into the refinery and marketing side of the oil business as well. It was the last stage in a chain that began with his first strike on the Nancy Taylor allotment.

His career seems to have been divided into certain easily recognizable spheres. First of all, he worked with his father in the oil business on a 30–70 per cent relationship. Then he began to build up his own oil interests, but for a long time continued to make his mark with his father. The story is well known of how he made his first oil strike himself; he was able to get a local banker, who was known to bid for the big oil companies, to act for him at an auction and bought the Nancy Taylor allotment for $500. The people on the rig had to beg him to keep out of the way as he was such a nuisance and so anxious to know if it was going to produce any oil. He waited on the platform while his friend, J. Carl Smith, came from the rig to report to him that it was flowing 30 barrels an hour – not, as Paul thought at first, 30 barrels a day. But, as he only had a 30 per cent stake in this strike, he did not make a great deal of money out of it.

Again and again we find him pointing out to his father and the board of directors where savings could be made and production more efficiently managed. Naturally, he received the treatment usually meted out to young men in a hurry. His father left $10 million dollars in his will, but the bulk went to his wife and only $500,000 to his son, Paul. It was not until after considerable struggling with his aged mother that he was able to persuade her to let him use her share to buy important assets during the Depression.

Late in life, when his businesses seemed to be comfortably established and not in need of further expansion, we find him entering the Alaska oil rush and bidding vast sums for stakes there in a possible strike which, needless to say, proved successful. Thus he was still prepared to lay out money in a speculative way to continue the dynamic expansion of his enormous empire although he told me that he was 'running scared' for fear that a man of his age might lose his judgement.

Long acquaintance with Getty convinced me that he would have made a very good dictator in the political field. At lunch at my club on St James's Street one afternoon with George Churchill, another avid collector, Getty showed himself to have great interest in dictators. He quoted a speech of Adolph Hitler at some length on the subject of settling issues by force. Hitler had been asked to settle some issue by peaceful negotiations and had replied: 'No great issue has ever been settled by peaceful negotiations. The question of the southern states of America was settled by force; the question of southern Ireland was settled by force.' Getty spoke with relish, and one could see the ruthlessness underlying his mask-like countenance gleaming forth.

| Will of George F. Getty | |
|---|---|
| Probate agreed at | $10,867,747 |
| Inheritance Tax | 1,300,000 |
| | 9,567,747 |
| and administrative | $8,419,000 to be distributed after bequests |
| To Paul Getty outright | $ 500,000 |
| To George F. Getty II in trust | 300,000 |
| To relatives and friends | 47,000 |
| To Christian Science Publishing Society | 1,000 |
| To Sarah Getty | $7,600,000 |

The public image of him as a miser sitting gloomily on his pile of wealth was surely surprising in a man who, by any standards, was one of the greatest public benefactors of this or any age. The fact that he had chosen the arts as his medium for philanthropy made this all the more unselfish as he was sharing, and in many cases giving away, treasures he would have enjoyed having in front of him. He spent $17 million on his museum in Malibu, California, and arranged for an endowment yielding $1 million a year to keep it open free to the public. What more could a man do to earn the title of a generous public benefactor?

Many people thought he was sad because his face was set as if in stone, but he had a great sense of humour and laughed frequently and spontaneously as the tapes bear witness. He showed me a letter – typical of many sent to millionaires in the hope of enlisting their financial support – addressed to his late son, George.

Dear George,
I know you are always interested in looking for opportunities for investment. I don't know if you would be interested in this, but I thought I would bring it to your attention because it could be a real 'sleeper' in making a lot of money with very little investment.

There is a developing market in the USA for cat skins, and a group of us are considering investing in a large cat ranch near Homosil, Mexico. It is our purpose to start rather small, about 1,000,000 cats, each cat averages about 12 kittens each year and skins can be sold for about twenty cents for the white ones and up to forty cents for the black. This will give us 12,000,000 per year to sell at an average price of around 32 cents, making a revenue about $3,000,000 a year. This really averages out at $10,000 a day excluding Sundays and holidays. A good Mexican cat man can skin about 50 cats per day at a wage of $3.15. It will only take 663 men to operate the ranch so that the net profit would be over $8,200 a day. Now, the cats will be fed on rats exclusively. Rats multiply four times as fast as cats. We would start a rat ranch adjacent to our cat farm. If we start with a million rats, we would have four rats per cat per day. The rats will be fed on the carcasses of the cats that we skin. This will give each rat a quarter of a cat.

You can see by this that this business is a keen operation, self-supporting and really automatic throughout. The cats will eat the rats, and the rats will eat the cats, and we will get the skins. Let me know if you are interested. As you can imagine, I am rather particular who I want to get into this and want the fewest investors possible. Eventually, it is my hope to cross the cats with snakes for they will skin themselves twice a year. This will save the labour costs of skinning as well, and give me two skins for one cat. Also, don't overlook the fertilizer market for the droppings. May I hear from you at your earliest opportunity?

Sincerely,<br>
H. Gregory Norton

One segment of his life which would seem to have contributed nothing to his success, but rather, caused him pain and anxiety, were his five marriages. My own impression is that none of them were significant to his career in a positive sense and were not supportive when their support would have been useful. One thing they did all have in common was that they did not marry him for his money. He had such a chip on his shoulder about this that he never seems to have thought of marrying a woman who would have been interested in his money.

Therefore, they were always interested in him and wanted his individual attention. It seems to me that the sort of person he should have married was a sex bomb who would have been delighted to say to him: 'Well, Paul, you keep making money; work as hard as you like; work sixteen hours a day, eighteen hours if you like. As long as you keep bringing home the money, I will keep the home fires burning and burning.' Instead, we find a series of women who were quite unaware of the significance of the person they were identified with, who complained in loud tones that they were neglected, and in the end sought a divorce or went back to mother.

I am in little doubt that the great love of his life was Marguerite Tallasou, the woman with whom he fell in love in Asia Minor in 1913 during his trip around Russia, Europe, and the Eastern Mediterranean. She was French and the wife of the Russian Consul-General for Asia Minor stationed in Brusa. When Paul was talking about her sixty years later, there were tears in his eyes (page 76). Obviously, he regretted deeply that he had never been able to make contact with her again as the result of the First World War and the Revolution

## Conclusion

in Russia. It may well be that, because of this romance, which ended on such an inconclusive note and in such dramatic circumstances, he was never again able to find himself wholly in love to the extent of dedicating himself to any particular woman.

He was quite conscious of the fact that he neglected his wives in the sense that he placed them second to business interests. Without them, of course, he would not have been able to produce a family and have the sons who bore his name. There was some unhappiness there, too, with one son dying young under anaesthetic during a minor operation and his eldest son dying in California in 1973 in mysterious circumstances. His relations with his son Paul were strained already when his grandson by him was kidnapped. No doubt publishers and critics alike would have wished him to unburden himself more in the tapes on the subject of his grandson's kidnapping. Actually, he described more of his part in that affair in his conversations with me than he had been prepared to say to newspaper reporters and others. Not for him the howls of anguish of an enraged grandparent; rather the decision to play it safe for the sake of his other grandchildren. His feelings were bitter enough when he said: 'Kidnapping is an unsavoury topic.' Yet, one must remember that when he talked of the 'company' doing something, he owned 64 per cent of the shares!

I think Paul spoke the truth when he said he would have made a better boyfriend than husband; and the girlfriends he had at different times in his life appeared to have more arresting personalities than the women he actually married. He agreed rather wryly with me when I said: 'The difference between marriage and a love affair is that in marriage you are remanded in custody; whereas in an affair you are only remanded on bail.'

I believe that one reason Getty settled down in England was that he had been received better there than in America or Italy. In America, there was a curiously persistent belief that Getty was a Jew. Even if this had been true, it would not have mattered in England or France, where we readily absorb our Rothschilds and Disraelis if they are interesting enough. To the Americans, even if correctly informed that he was of Irish and Scottish blood, his father and mother lived such quiet lives as to be considered almost bourgeois. It would have to be Paul Getty's grandchildren who would take their place beside the Duponts, Vanderbilts, and Rockefellers in America's 'moneytocracy' (to coin a word, if I may).

When he established a refinery at Gaeta, he bought the ruined sixteenth-century Palazzo Vecchio adjoining the Odescalchi Castle north of Rome and restored it. He naturally wanted to be accepted in Rome on his own merits – a man of civilized tastes with a painstakingly acquired collection, a linguist, and a gentleman. He was put up for membership in a leading Rome club, but to his astonishment and mortification, was blackballed. The Italians of the *dolce vita* set mistook him for just another self-made millionaire whom they were always turning away from their doors.

In England, on the other hand, Getty was always accepted by the prominent families. It was not that he was consciously snobbish, but simply that he enjoyed the company of significant people. The guest list for his first dance at Sutton reads like a handful of pages out of Debrett. He was liked in England. His immense erudition and taste in the arts endeared him to the more influential English. Withal, he was quiet and unassuming, the very antithesis of the qualities often attributed – and often wrongly attributed by the English – to his compatriots. The English took him to their bosom and would have been sorry had he kept to his often-announced intention of retiring to spend his last days in the sunshine of his native California, within a few steps of the great museum he had built, to be reunited with his art treasures in the evening of his life. As it was, he died in England in that most English of all English houses, Sutton Place. He lived a long, extraordinary life; his achievements remarkable, his legacy enormous, his secret, 'genius'.

# APPENDIX I

4th October, 1973.

Mr. Somerset de Chair,
St. Osyth's Priory,
St. Osyth,
Essex.

Dear Somerset,

You telephoned me the other day and told me that you had been asked to do a biography of me by your publisher, and was I willing to co-operate with you by giving you a reasonable amount of my time and access by you to a reasonable amount of information about me. I am willing that you should do the biography of me provided that it is agreed by you and your publisher that (1) prior to any publication of the biography of me, you will submit your final manuscript including the title selected for my approval (2) you will make such changes, deletions or additions to the manuscript and the title as I may request and (3) no publication of the manuscript shall be made unless my written approval of the final manuscript is first obtained. You and your publisher further agree to my sole determination of what is reasonable in regard to my time and the material furnished by me shall be at my sole discretion, not yours.

I am thinking of giving you approximately one hour a business day for four or five weeks and that will constitute my present determination of a reasonable amount of time. As to material, I will give you permission to examine at Sutton

Place, my diary, such albums and scrapbooks as I have here and personal letters of mine from my files here, to the extent I may make them available to you, subject to my sole discretion as to whether they may be published or not.

On further thought, I cannot approve the pun title, "The Life and Dimes of J. Paul Getty". It is, of course, your responsibility and that of your publisher to ensure that there is no conflict between the title you select and any other title.

It is understood that I am not financially responsible for anything connected with your proposed book and by the same token, you are free, so far as I am concerned, to whatever may accrue to you from the writing of the proposed book. You and your publisher agree to hold me harmless and to indemnify me against all expense or liability of any kind arising from the writing and publication of the proposed book.

It is understood that there is no responsibility on my behalf to furnish you any more of my time than the above mentioned one hour a business day for four or five weeks. It is understood that if due to physical reasons or incapacity, either you or I find it impossible to carry out your plan as set forth herein, my staff would co-operate with you in my absence, subject to my approval or that of my personal representative or executor as provided above, and your literary executor, subject to this agreement, would finish your book, in your absence or incapacity.

If the foregoing is acceptable to you and your publisher, I would appreciate your returning to me a copy of this letter signed by you and your publisher.

Sincerely,

*J Paul Getty*

J. PAUL GETTY

# APPENDIX II

Sutton Place, Guildford

After the extraordinary business of the faked biography of Howard Hughes by Clifford Irving, I feel it desirable to state unequivocally that I have read and approved every word of the present book. The recordings of the conversations between me and Somerset de Chair were indeed made at Sutton with my knowledge and approval in October and November 1973 and February 1974. My only criticism is that the portrait of me presented by Somerset de Chair is too flattering, but I am quite prepared to be judged by it as it stands.

Signed

*Paul Getty*

# INDEX

*Note* All subject references are to John Paul Getty unless specifically indicated otherwise.

Abdication (1936) 79–81
Acquisitive nature 148
Agnew, Colin 151
Aircraft business 30, 137
   *see also* Spartan Aircraft Company
Albert, Prince Consort 79
Alexander the Great 22
Alps, description of 125
Anastasia, Grand Duchess 84
Arab-Israeli war (1973) 109–10, 112
Arabian oil 148–9
Arabic, speaking 107
Ardabil carpet 53, 149
   description of 134–5
Art
   auctions 150–1
   collection 37–40, 47–53
Artists 41
Ashby, Allene (second wife) 67
Assassination, Grand Duke Ferdinand (1914) 110
Athens Museum 50, 52
Attorneys *see* lawyers
Autobahn 127

Baldwin, S. 81
Baptism as Jean Paul 59
Bauer collection 132–3
Bedford, Duke of 83
Belgium, touring 132
Bell, Doctor 60
Blucher, Marshal 142
*Blue Boy* (Gainsborough) 48
Bow, Clara 69
Boymans Museum 132
British Museum 48, 50
British Petroleum (BP) 33
Brockhurst, Gerald 131, 150
Bronze deer 50

Browning, Robert 108
Bullimore (butler) 37
Business/businessmen 17–19, 22–36
   Getty on 13–14, 22, 24–5, 28, 32, 33, 35, 98

Caesar, Augustus 16
Caesar, Julius 16, 44
Calpurnius Piso 44
Canton, plague in 88
Carcaci, Duchess of (Castilla, Marie née Millington-Drake) 74, 146–7
Carlin console 132
Castilla *see* Carcaci
Cats, letter about 157–8
Caviar 85–6
Chace, Mr (detective) 5
Chance, Peter 50
'Charioteer of Delphi' 52
Cheremetiev, Count 95–6, 135
China 93, 110–11
   visit to 87, 88
Churchill, George, 50, 156
Churchill, W. S. 16, 40, 79, 117
Cicero 44
Clairvoyant 147
Clifford, Alice 75
Clifford, Bede 75
Coca-Cola Company 21
Collections
   beginnings 40
   contents of 48
   interests 53–4
   marbles 123
Collector, Getty as 38–9
Communism 110–11
Conservatism 15
Constable-Maxwell, Jeanette 74
Cooper, Diana 85

Cretan bull 50
Curie, Mme 148
Czechoslovakia, occupation of 131

Davies, Marion 6
Debts (in 1938) 134
Dempsey, Jack 101–2, 104
'Diana and Actaeon' (Titian) 54
Diaries
  1904 122–4
  1938 124–35
  1939 135–7
Dictators 117
  interest in 14, 156
Diplomats 116
Divorces 69, 72–3
Dixon, Jean (clairvoyant) 45–6
Drillers 64
Drinking habits 144
Durrell, Lawrence 74
Duveen, Lord 53, 134

Earthquakes, threat of 47
Edward, Duke of Windsor *see* Windsor
Edward, Prince of Wales 77
Edward VII, King 79
Elizabeth, Grand Duchess of Russia 84
Elizabeth (Queen Mother) 82
England, settled in 8, 159–60
*Europe in the Eighteenth Century* (book) 120–1
Environment and pollution, Getty on 114–15
Exxon Company 33

Fabre collection 132
Farlie House (Scotland) 74
Family background 56–9, 60
Farrish (head of Standard Oil of New Jersey) 26
Faust 128
Ferdinand, Grand Duke 110
First World War 94
  enlisting 31
Fixed price contracts, avoiding 29–30

Ford, Henry 37
France 110
  touring 132–4
Frick collection 48, 135
Froshlin, Hansi 128
Fruit for breakfast 128–30
Furness, Thelma 80

Gaeta (Italy), refinery 7, 160
Gainsborough 48
Gambling, Getty on 16
Geology, study of 61–2
Geordie, Duke of Sutherland 7, 10, 11, 75
George F. Getty Inc. 17, 154
  debts 134
George III, King 48
George V, King 79
George VI, King 79
Germany
  advance into Poland 136
  autobahn 127
  return to (1939) 136–7
  touring 126–30
Getty, George (Civil War general) 152–3
Getty, George F. (father) 17, 22, 57, 59, 60
  book written with 120
  influence of 97
  will of 153
Getty, James (founder of Gettysburg) 56–7
Getty, Jeanette (first wife) 66
Getty, John (first American ancestor) 56, 152
Getty, John (grandfather) 57
Getty, John Paul
  on abdication (1936) 81
  approval for book 1, 161
  on artists 41
  at 81 years 1–2
  on boxing 102–3
  on business/businessmen 13–14, 22, 24–5, 28, 32, 33, 35, 98
  on communism 110
  on Duke of Windsor 78
  on early romance 75

## Index

    on environment and pollution 114–15
    on family 56–9
    on gambling 16
    on inequality 20
    on investment 34–5
    on judgement 143
    on judges 117
    on lawyers 73, 114, 146
    on marriage 65–6, 70, 72
    on Napoleon and Hitler 14
    on Nixon tapes 118
    on oil shortage 115–16
    on opportunity 26
    on politics 116–19
    on relaxation 106
    on religion 61
    on smoking 119
    starting out 24
    on subordinates 143
    on wealth 19–21
    on weightlifting 145
    on work 12–13
    on writers and writing 120
Getty, John (uncle) 58
Getty, Martha Ann Wiley (grandmother) 57
Getty Oil, ownership of 23
Getty, Ronny (son) 125–6, 132–3
Getty, Sarah C. (mother) 64
Getty, Senator William Reid 57, 153
Goebbels, J. 137
Göring, H. 137
Goudstikker 131
Grand Army of the Republic 112
Grouchy, Marshal 142
Guest, Mrs 40

Harriman, A. 119
Hays, Lansing 1
Hearst, W. R. 5–7, 144
Hecht, David 114, 127, 128, 135
Helmle, Fini (third wife) 68, 125–6, 133
Herculaneum
    villa 42–3, 55
    architecture 44–5
'Hercules' (statue) 54

'Hermes' (Praxiteles) 52–3
Hermitage (Leningrad) 95
Hess, R. 137
Hewins, Ralph 152
Hillington collection 135
*History of the Oil Business* (book) 60, 97–8, 120
Hitler, A. 14–16, 130, 136, 137, 156
Honolulu 87, 89
Hopkins, J. 26
Horseracing 144
*How to be Rich* (book) 17, 119
Hughes, Howard 1
Huntingdon Library (California) 48

Inchcape, Lord 74
Inequality, Getty on 20
Insights
    into environment and pollution 114–15
    into lawyers 114
    into oil shortage 115–16
    into politics 116–19
    into writing and writers 120–1
Investment, Getty on 34–5
Irving, Clifford 1
Istanbul 76
Italy, visit to 135–6, 137

Japan 87–9
*The Journey from Corinth* (book) 42
*The Joys of Collecting* (book) 120, 149
Judgement, Getty on 143
Judges, Getty on 117
Juliet, Marchioness of Bristol 106, 142, 144

Kennedy family 119
Ketchell, Stanley 103
Kidnapping of grandson 4–5, 159
Kipling, Rudyard 120
Kitson, Penelope 65, 74, 75

Lacroix, Leon 150
Lawyers, Getty on 73, 114, 146
Leander, Zarah 130

Leeds Castle, Kent 6
Legh-Jones, Sir John 154
Letter/s
  begging, to George Getty (son) 157–8
  to Getty 137–41
  writing 98–9
LeVane, Ethel 42, 120
Lions 106
London 75
  transfer to 9–10
Lynch, Teddy (fifth wife) 68, 70, 126, 127, 130–1, 133, 136, 137

*The Madonna of Loreto* 150–2
Magazines 99–100
Making money 153
Malibu Museum 7, 39, 41–2
  destruction prophesied 45–6
  endowment 156–7
  tapestries 91
Manila 88–9
Marble, qualities of 49
Marbles, collecting 99, 123
Margaret, Duchess of Argyll 65
Margarete *see* Faust
Marriage 65, 158–9
  Ashby, Allene (second wife) 67
  Getty on 65–6, 70, 72
  Helmle, Fini (third wife) 68, 125–6, 133
  Jeanette (first wife) 66
  Lynch, Teddy (fifth wife) 68, 70, 126, 127, 130–1, 133, 136, 137
  Rork, Ann (fourth wife) 68–9
'Marten Looten' (Rembrandt) 41, 132
Mauritshuis Museum 132
Mellon, Andrew 135
Menzies, Alistair 77
Michelangelo 49, 53
Middle East 31
Milland, Ray 72
Millington-Drake, Effie 74
Millington-Drake, Eugene (ambassador to Uruguay) 74
Mission Corporation 26, 78, 154

Mitchell (president of the National City Bank) 118
Mitchell, Charley 114
Monarchy 80–2
*Moses* (Michelangelo) 49
Muller, Maris 128
Murietta, Joaquin 144, 145
Museums
  Athens 50, 52
  Boymans 132
  British 48, 50
  Los Angeles 149
  Malibu 7, 39, 41–3, 45–6
  Mauritshuis 132
  Prado 51
  Rijksmuseum 131
  Victoria and Albert 48, 53
Mussolini, B. 16
*My Life and Fortunes* (book) 59

Nancy Taylor allotment 155
Napoleon, Emperor 14–16, 89, 101, 152
National Gallery 41, 48, 152
Netherlands, touring 130–2
Neuerburg, Professor 44
Nixon, Richard 143
  Watergate tapes 117–18
North, Lord 48
Norton, H. Gregory, letter to George Getty 158

Odescalchi castle (Italy) 7, 160
Offshore drilling 62
Oil
  Alaskan 155
  Arabian investments in 148–9
  in Baku 91
  beginnings in 24, 61
  drilling 62–4, 63–4
    offshore 63
  father's beginnings in 61
  first strike 155
  shortage 115–16
  refining, blocked 154
Opportunity, Getty on 26
Ostankino Palace 95–6
Oxford 77, 90

*Index*

Paris 75
Partridge, Frank 37
Posta Vecchio (villa) 7, 160
Peaches *see* fruit
Pelton, Miss (childhood teacher) 122–3
Penn, William 56–7
Personal papers 122–41
Pets
  Gip 105, 106, 122
  Shaun 106
Philip, Duke of Edinburgh 82
Philip of Macedon 22
Phillips, Donald K. 139–41
Pierre Hotel 128, 133, 135, 154
Pierre Marques Hotel 154
Pine, Senator 58
Poland, invasion by Germany 136
Political involvement 119
Pompeii 42, 43–4
  visit to 137
Prado Museum 51
Prices, in Paris (1938) 133

Raphael, *The Madonna of Loreto* 150–2
Rasputin 83, 85
Relaxation, Getty on 106
Religion, Getty on 61
Rembrandt 20, 41, 47, 49, 54, 84–5, 132
Revolcadero beach (Mexico) 153–4
Ribbentrop, J. von 137
Rijksmuseum 131
Rockefeller, John D. 13–14
Rockefeller, John D. Jr 26
Rockefeller, N. 119
Romances, early 75–6
Rome, house near 8
Roosevelt, F. D. 16
Rork, Ann (fourth wife) 68–9
Roughneck 63
Roustabout 63
Rubenstein, Arthur 98
Rugs, purchase of 135
  *see also* Ardabil carpet
Ruisdael, painting by 38
Russia 85, 90, 92, 94–6, 110–11
  prince of 91–3
  revolution 84
  tapestries 90–1
Ryan, Thomas Fortune 40

St Bartholomew (painting of, by Rembrandt) 47
St Donat's Castle 6–7
San Simeon 6, 26
*Saturday Evening Post* 99–100, 124, 148
Saud, King ibn 107
Savonnerie 132, 133
Scharf, Dr A. 151–2
Schulein, Mr 131
Secretaire, purchased from Ball 133
Second World War 28–9, 153
  scare over 127–31
Seymour, Hal 153
Shares, transactions in 134
Shell Oil Company 33, 154–5
Simon, Norton 91
Simpson, E. 80
Simpson, Mrs W. 79, 80–1
Sinclair, Harry 27
Smith, J. Carl 155
Smoking, quitting 119
'Soldier of Marathon' 50
Spartan Aircraft Company 28–9, 71, 153
Stalin, S. 16
Standard Oil of Indiana 13
Standard Oil of New Jersey 26, 33
Starr, Henry 59–60
Steuer, Max 118
Stewart, Colonel (president of Standard Oil of Indiana) 13–14
Subordinates, Getty on 143
Sun Yat Sen, Dr 89
Susa, Charlotte 129, 134, 136, 137
Sutton Place 1–2, 160
  acquisition of 8–11
  origins of 11
  visit by Hearst 7
  and women friends 75
Swimming 87–8

Switzerland, touring 125–6
Synagogue, visit to 135

Talents 20–1
Tallasou, Marguerite 75–6, 158
Tapestries 91
Teagle, (head of Standard Oil of New Jersey) 26
Teissier, Marie 75, 84, 90
Thyssen, Heini 145–6
Tidewater Associated 6, 131, 149, 154
  merged 23, 25–6
Tier, Carol 36
Titian 54
'Titus' (Rembrandt) 54
Touring
  Belgium 132
  France 132–4
  Germany 126–30
  Netherlands 130–2
  Switzerland 125–6
Travels
  in Europe 90–4
  in Far East 87–90

Venus, crouching 51–2
Vermeer 85
Victoria and Albert Museum 48, 53
Victoria, Queen 79
Vienna 68, 93–4
von Alvensleben, Baroness 106

Walker, Elisha 27
War
  Arab-Israeli (1973) 109–10
  possibilities of 111
Watergate 117–18
*The Waterloo Campaign* (book) 142
Watson, Mr and Mrs 36
Wealth, Getty on 19–21
Weber (Swiss engineer) 43
Weightlifting 106, 145–6
Westminster, Duke of 37
Weston, Dorothy A. 86
Weston, Francis 11
Weston, Sir Richard 11
Windsor, Duke of 78
  abdication 79–80
Wives *see* marriage
Woburn 83
Wolfson, Sir I. 54
de Woolf Grady, Karen 137–9
Work, Getty on 12–13
Writers/writing 120–1

Yusupov, Prince 83–4, 85

Zionism 112